# Quaternions

by

Dennis Morris

© Dennis Morris

All Rights Reserved

Published by: Abane & Right

31/32 Long Row

Port Mulgrave, Saltburn

TS13 5LF

United Kingdom

01947 840707

dennis355@btinternet.com

November 2015

# Contents

Introduction ..................................................... 1
An Overview of Numbers ................................... 3
    The division algebra axioms: ............................ 4
    What about multiplicative commutativity?: ........ 7
    The 1-dimensional real numbers: ....................... 8
    The 2-dimensional complex numbers: ................ 8
    Rumours: ............................................................ 9
    A scuppering of rumours: ................................ 11
    Derivation of the complex numbers: ................ 12
    Summary: ......................................................... 15
Derivation of the Quaternions .............................. 17
    Forming the division algebras: ........................ 26
    Emergent expectation distance functions: ........ 30
    Square roots of real numbers: .......................... 32
Conventional Notation ........................................ 34
Quaternion Rotation ........................................... 36
    Connections between trig. functions: ............... 38
    The polar form of the quaternions: ................... 39
    No rotation axis: .............................................. 40
    Double cover rotation: ..................................... 41

## Contents

    Violation of parity perhaps: ................................43

    Quaternion rotation as a continuous group: ......43

    Restricted rotation in the quaternion space: ......45

    Non-commutative quaternion rotation: .............46

    A non-commutative form of rotation: ...............49

    Clifford algebras and double cover: ..................54

    Quaternion multiplication revisited: .................55

Non-commutative Differentiation ............................58

    Differential operators: .......................................62

    The Anti-quaternions: ......................................67

    Comments: .......................................................67

Double Differentiation .............................................69

Connections to Clifford Algebras ............................71

Commutation Relations and Lie Groups ..................73

The Quaternion Trig. Functions ...............................79

    Second differentials: ........................................82

    Third differentials: ...........................................83

    Fourth differentials: .........................................84

    Orthogonality: ..................................................84

The Quaternion Inner Product .................................89

    The Euclidean inner product: ...........................89

    Trigonometric identities: ..................................92

The Spinors of Physicists .........................................95

- Spinors and division algebras: ..........................98
- The inner product: ...........................................102
- Anti-quaternions and anti-spinors: ..................107
- Interim summary: ............................................109
- Bilinear Covariants and the Dirac spinor: ........110
- Classification of spinors: ..................................113
- Postmethian comments: ...................................113

A Note on Quaternion Fourier Analysis .................115

Concluding Remarks................................................117

Other Books by the Same Author ..........................120

Index .......................................................................125

Chapter 1

# Introduction

Recent developments in theoretical physics have led to the view that a kind of 4-dimensional numbers called quaternions might have a central role to play in theoretical physics. This view is not a new view; it was the view of quaternions held by physicists in the latter half of the 19th century. However, during the 20th century, this view was cast aside. Remarkably, except for a few NASA engineers[1], quaternions have been ignored or put into dark corners throughout the whole of the 20th century. Only a brave few mathematicians and physicists have taken any interest in quaternions, and quaternions are usually absent from the curricula of mathematics and physics degrees. All this in spite of the fact that quaternions are a Clifford algebra holding double cover spinors and the $SU(2)$ commutation relations that are so central to quantum physics. Well! it seems they are back.

Since quaternions were first discovered in 1843, knowledge and understanding of them has been gained only spasmodically over that last 170 years. Today, we find different aspects of the quaternions scattered over many books but no single book that pulls together the different

---

[1] NASA engineers use quaternions to control the attitude of the space shuttle.

aspects of them that might be of interest to a physicist. Hence the need for this book.

In this book, your author pulls together into a single volume our understanding and knowledge of quaternions. The book includes the non-commutative differentiation of quaternions and the non-commutative rotation of quaternions. This book also includes an illuminating chapter upon the spinors used in quantum physics.

The book is primarily a mathematics text with only a single chapter of the spinors so beloved by theoretical physicists, but your author hopes that physicists will read this book and that it will enable them to apply quaternions to their work. By pulling our understanding of quaternions together, your author hopes he has provided a stepping stone for the physicist. Your author hopes that he has contributed to the improvement of human knowledge and facilitated a deeper understanding of our universe.

Chapter 2

# An Overview of Numbers

Quaternions are a type of numbers. To understand quaternions, we need to understand numbers. Examples of types of numbers are the real numbers, $\mathbb{R}$, with which we have been accustomed since childhood and the Euclidean complex numbers, $\mathbb{C}$, with which we have been accustomed since about the start of a university degree.

Mathematicians have a technical name for types of numbers; they call them division algebras. Technically, a division algebra is a set of mathematical objects which satisfies the thirteen division algebra axioms. For completeness, we will list these axioms. This is the most technical part of this book. It is not essential to the understanding of quaternions to thoroughly understand the technicalities of the division algebra axioms. The physicist reader is advised to read this next section once and not to worry about it further.

None-the-less, like drinking enough beer to make you ill for a week, every human being should see the list of division algebra axioms at least once in their lifetime. You have not lived if you have not seen the division algebra axioms.

## The division algebra axioms:

The thirteen division algebra axioms are traditionally[2] taken to be:

F1: Addition is commutative

F2: Addition is associative

F3: Inclusion of an additive identity

F4: Inclusion of additive inverses*

F5: Additive closure

F6: Multiplicative closure

F7: Inclusion of multiplicative inverses

F8: Inclusion of multiplicative identity

F9: Multiplication is associative

F10: Multiplication is distributive over addition

F11: The multiplicative identity is not the additive identity

F12: The multiplicative product of the additive identity and every other element of the set is the additive identity

F13: There are no zero divisors

We give brief explanations of these axioms.

---

[2] There is some choice of which set of axioms we choose to define a division algebra. The choice is cosmetic, but the reader might meet other, but equivalent, sets of axioms elsewhere.

*An Overview of Numbers*

F1: The commutativity of addition is simply that:

$$a+b = b+a \qquad \forall a,b \qquad (2.1)$$

The $\forall$ sign means 'for all'. The $a,b$ following this sign is the list of elements to which the 'for all' sign refers.

F2: The associativity of addition is simply:

$$a+(b+c) = (a+b)+c \qquad \forall a,b,c \qquad (2.2)$$

F3: The additive identity is 0 such that:

$$a+0 = a \qquad \forall a \qquad (2.3)$$

F4: The inclusion of additive inverses means that, for every element of the set, there is a corresponding element which, when added to the first element produces zero. Roughly, we are including negative numbers in the set:

$$a+(-a) = 0 \qquad \forall a \qquad (2.4)$$

Within the list of axioms above, we have placed an asterisk upon this traditional axiom because we are going to change it. We are going to change it to, "for every element of the set, there is a corresponding element which, when added to the first element produces zero *except for the real part of the division algebra*". We will explain this change in more detail later.

F5: Additive closure means that, if two elements of the set are added together, the result is another element of the set. A number added to a number produces a number.

F6: Multiplicative closure means that, if two elements of the set are multiplied together, the result is another element of the set. A number multiplied by a number produces a number, or, if the reader prefers, a duck mated with a duck produces a duck and not a chicken. There are many objects in mathematics which do not have multiplicative closure; for examples, a $1 \times 4$ matrix multiplied by a $4 \times 1$ matrix produces a $1 \times 1$ matrix and two vectors 'dotted together' produce a number rather than a vector.

F7: The inclusion of multiplicative inverses means that, for every element of the set except zero, there is a corresponding element which, when multiplied by the first element, produces the multiplicative identity which is the number 1.

$$a \times \frac{1}{a} = 1 \qquad \forall a \text{ except } 0 \qquad (2.5)$$

F8: The multiplicative identity is the number 1 such that:

$$a \times 1 = a \qquad \forall a \qquad (2.6)$$

F9: Multiplication is associative means:

$$a \times (b \times c) = (a \times b) \times c \qquad \forall a,b,c \qquad (2.7)$$

The reader might be surprised to discover that there are mathematical objects, the 8-dimensional octonians for example, which are not associative.

F10: Multiplication is distributive over addition means:

$$a(b+c) = ab + ac \qquad \forall a,b,c \qquad (2.8)$$

F11: The multiplicative identity is unity, 1, and this is not equal to the additive identity which is zero, 0.

F12: The product of zero with any other member of the set is zero:

*An Overview of Numbers*

$$a \times 0 = 0 \qquad \forall a \qquad (2.9)$$

F13: There are no zero divisors means that, if the multiplicative product of two elements of the set is zero, then one or both of the two elements must be zero:

$$a \times b = 0 \Rightarrow a = 0 \text{ or } b = 0 \\ \text{ or } a = b = 0 \qquad (2.10)$$

There are mathematical constructions such as many of the Clifford algebras which do have zero divisors and are thus not division algebras (types of numbers)[3]. In general, if a mathematical construction contains the square root of plus unity, $\hat{r} = \sqrt{+1}$, then that mathematical construction contains zero divisors; we have:

$$\left(1 + \sqrt{+1}\right)\left(1 - \sqrt{+1}\right) = 1 - \sqrt{+1} + \sqrt{+1} - \sqrt{+1}\sqrt{+1} = 0 \qquad (2.11)$$

Such zero divisors can be avoided by taking the polar form of the mathematical construction – more on this later.

## What about multiplicative commutativity?:

The astute reader will have noticed that the division algebra axioms do not include the requirement of multiplicative commutativity. We do not need $a \times b = b \times a$. If we had included multiplicative commutativity as an axiom in the above list, then we would have had the axioms of an algebraic field. An algebraic field is just a multiplicatively commutative division algebra, and a division algebra is just a, not necessarily but could be, multiplicatively non-

---

[3] But see: Dennis Morris: The Naked Spinor

commutative algebraic field. In practice, the terms 'non-commutative algebraic field' and 'division algebra' are used interchangeably as are the terms 'commutative division algebra' and 'algebraic field'.

The quaternions are famously multiplicatively non-commutative, of course. The Euclidean complex numbers and the real numbers are equally famously multiplicatively commutative. Since all division algebras are additively commutative, when we use the words 'commutative' and 'non-commutative', we will always mean multiplicatively so even though we omit the word 'multiplicatively'.

Well! see how much your life has been enhanced by this viewing of the axioms of a division algebra. Like drinking enough beer to make you ill for a week, you have learned something well worth knowing. You will have much to speak of in the pub tonight.


The 1-dimensional real numbers:
Right, that's the technical stuff out of the way, and so we can now get started on the guts of this book.

The reader will be familiar with the 1-dimensional real numbers, and so we will not dwell upon them.


The 2-dimensional complex numbers:
The reader will probably be familiar with the 2-dimensional Euclidean complex numbers, $\mathbb{C}$, which she might know as the 'imaginary numbers' or as simply the 'complex numbers'. We will dwell upon these numbers as they are an

*An Overview of Numbers*

essential stepping stone to the quaternions. The Euclidean complex numbers are of the 2-dimensional form[4]:

$$a + \hat{i}b \quad : \quad \hat{i} = \sqrt{-1} \qquad (2.12)$$

In a manner analogous to the way we think of the real numbers, $\mathbb{R}$, as positions upon a line, that is as positions in 1-dimensional space, we can think of the 2-dimensional complex numbers as positions in a 2-dimensional Euclidean plane.

Rumours:

There is a rumour going around that the Euclidean complex numbers are 'an algebraic extension of the real numbers based on a monic minimum polynomial'. This rumour has been circulating for over 400 years. The rumour is not untrue, but it is very misleading. The idea of the rumour is that we begin with the real numbers and we write the polynomial:

$$x^2 + 1 = 0 \qquad (2.13)$$

This is a monic polynomial because the coefficient of the highest power of the variable $x$ is one (one – mono). It is a minimal polynomial because it cannot be factorised using only real numbers. We see that, because this polynomial, (2.13), cannot be factorised into linear factors of the form $(x \pm \alpha) : \alpha \in \mathbb{R}$, this polynomial cannot be solved using only the real numbers. To solve this polynomial, (2.13), we have to extend (that's where the 'extension' comes from) the

---

[4] Note: It is unconventional to put a carat over the letter $i$. Your author does it to be consistent with later notation.

*Quaternions*

real numbers into two dimensions by introducing the 'new' number $\hat{i} = \sqrt{-1}$; this gives:

$$(x+\hat{i})(x-\hat{i}) = x^2 + 1 = 0 \qquad (2.14)$$

and we have a solution to the polynomial.

Although the idea of extending the real numbers into higher dimensions using a polynomial which cannot be factorised into linear factors 'works' for the Euclidean complex numbers, it misled mathematicians for some 400 years. During that 400 years, mathematicians sought 3-dimensional complex numbers by seeking monic minimum $\mathbb{C}$ polynomials (polynomials that cannot be factorised within the $\mathbb{C}$ algebra). Such a minimal polynomial would necessitate the extension of the 2-dimensional Euclidean complex numbers into three dimensions. The search was ended in 1799 when the mathematician Johann Carl Friedrich Gauss (1777-1855), aged only 19 at the time, proved that there are no polynomials in Euclidean complex variables which cannot be factorised into linear factors. This was Gauss's doctoral thesis; he produced three other proofs of the same fact throughout his lifetime. The phenomenon of being able to linearly factorise all polynomials written in Euclidean complex variables using only Euclidean complex numbers is called the 'algebraic closure' of the Euclidean complex numbers.

Well! that was that. Clearly there were no 3-dimensional complex numbers and, by implication, no 4-dimensional numbers nor numbers of any dimension higher than two. That might have been that, but, in 1843, William Rowan Hamilton (1805-1865) discovered the 4-dimensional quaternions. John T. Graves then discovered the 8-

*An Overview of Numbers*

dimensional octonians, now called Cayley numbers[5] two months later, and in, 1848, James Cockle (1819-1895) discovered another type of 2-dimensional complex numbers, the hyperbolic complex numbers. Clearly, in the eyes of the 'algebraic extensionists', the 4-dimensional quaternions could not exist; except, they did exist.

The response of most of mathematical academia to the discoveries of Hamilton, Graves, and Cockle were to accept Graves discovery but to call the 8-dimensional octonians after the mathematician Arthur Cayley (1821-1895), to 'forget' all about Cockle's discovery, and to declare that Hamilton's quaternions, although very interesting and perhaps useful, were not 'proper' numbers because they were multiplicatively non-commutative and not a 'proper' extension of the Euclidean complex numbers. Let us face it; everyone accepts that Graves's octonians are not 'proper' numbers, and so, perhaps, quaternions are not 'proper' numbers.[6]

There is no monic minimum polynomial which gives rise to the quaternions.

A scuppering of rumours:
The 2-dimensional complex numbers, both types, derive from the order 2 finite group $C_2$. The quaternions derive from the order 4 finite group $C_2 \times C_2$. Any concept of

---

[5] Cayley numbers are non-associative and therefore not 'proper numbers'. They are unfairly named after Cayley.
[6] The reader should not think of mathematicians as being above the human frailties of obstinacy, stupidity, vanity, and academic myopia.

'algebraic extension based on a monic minimum polynomial' is, in your author's opinion, best forgotten.

## Derivation of the complex numbers:

We begin with the finite group $C_2$. We could think of $C_2$ as just the two numbers $\{+1,-1\}$. However we multiply these two numbers together, we get one or other of the numbers (multiplicative closure). One of these numbers, $+1$, is the multiplicative identity, and both numbers are their own multiplicative inverses:

$$+1 \times +1 = +1 \quad \& \quad -1 \times -1 = +1 \quad (2.15)$$

Instead of thinking of $C_2$ as the two numbers $\{+1,-1\}$, we will write the group $C_2$ as two $2 \times 2$ permutation matrices:

$$\begin{bmatrix} 1 & 0 \\ 0 & 1 \end{bmatrix} \quad \& \quad \begin{bmatrix} 0 & 1 \\ 1 & 0 \end{bmatrix} \quad (2.16)$$

With a little effort, the reader will see that these two matrices have exactly the same multiplicative relations as the numbers $\{+1,-1\}$. The reader might notice that:

$$\begin{bmatrix} 0 & 1 \\ 1 & 0 \end{bmatrix}^2 = \begin{bmatrix} 1 & 0 \\ 0 & 1 \end{bmatrix} \quad (2.17)$$

The reader might think, 'we have a square root of plus unity, and, from earlier, we will thus have zero divisors'. Well spotted, but you've jumped the gun.

We replace the 1's with real variables and add the two matrices:

$$\begin{bmatrix} t & 0 \\ 0 & t \end{bmatrix} + \begin{bmatrix} 0 & z \\ z & 0 \end{bmatrix} = \begin{bmatrix} t & z \\ z & t \end{bmatrix}$$

$$\det\left(\begin{bmatrix} t & z \\ z & t \end{bmatrix}\right) = t^2 - z^2$$

(2.18)

Testing the resultant matrix against the axioms of a division algebra listed above, we soon discover that, because the resultant matrix has a determinant that can be zero, not every matrix will have a multiplicative inverse and so this resultant matrix is not a division algebra. There are two ways forward. The first way forward is to take the exponential of the matrix and thereby be rid of the singular matrices:

$$\mathbb{S} = \exp\left(\begin{bmatrix} t & z \\ z & t \end{bmatrix}\right) = \begin{bmatrix} e^t & 0 \\ 0 & e^t \end{bmatrix} \begin{bmatrix} \cosh z & \sinh z \\ \sinh z & \cosh z \end{bmatrix}$$

(2.19)

This is the 2-dimensional hyperbolic complex numbers discovered by Cockle. These numbers have no additive inverses on the real axis, but they satisfy all the other division algebra axioms (we are rid of the zero divisors). The rotation matrix of the hyperbolic complex numbers, (2.19), is the Lorentz boost of special relativity (rotation in 2-dimensional space-time). Because the real axis is time, and we cannot travel backwards in time, we will not worry about additive inverses on the real axis. We have:

*Quaternions*

$$\begin{bmatrix} \cosh \chi & \sinh \chi \\ \sinh \chi & \cosh \chi \end{bmatrix} = \begin{bmatrix} \gamma & v\gamma \\ v\gamma & \gamma \end{bmatrix} \quad (2.20)$$

Where:

$$\gamma = \frac{1}{\sqrt{1 - \frac{v^2}{c^2}}} \quad (2.21)$$

The whole of special relativity is within the hyperbolic complex numbers[7]. Since this is a book about quaternions, we will be concerned with the hyperbolic complex numbers no longer.

The second way forward is to be rid of the singular matrices by throwing a minus sign in to the matrix:

$$\begin{aligned} \mathbb{C} = \begin{bmatrix} a & b \\ -b & a \end{bmatrix} &\equiv a + \hat{i}b \\ \det\left(\begin{bmatrix} a & b \\ -b & a \end{bmatrix}\right) &= a^2 + b^2 \neq 0 \end{aligned} \quad (2.22)$$

This is the Euclidean complex numbers. The $2 \times 2$ matrix is just another way of writing these numbers – it is, in your author's opinion, a much better way of writing the Euclidean complex numbers.

Taking the exponential of the Euclidean complex numbers matrix gives the rotation matrix of 2-dimensional Euclidean space:

---

[7] See: Dennis Morris: Empty Space is Amazing Stuff.

*An Overview of Numbers*

$$\exp\left(\begin{bmatrix} a & b \\ -b & a \end{bmatrix}\right) = \begin{bmatrix} e^a & 0 \\ 0 & e^a \end{bmatrix} \begin{bmatrix} \cos b & \sin b \\ -\sin b & \cos b \end{bmatrix} \quad (2.23)$$

So, both types of 2-dimensional complex numbers derive from the finite group $C_2$. Very nice, but what about the 1-dimensional real numbers? They derive from the finite group $C_1$ which we can think of as just the number one or as a $1\times1$ permutation matrix:

$$\begin{aligned} [1] &\to [a] \\ \exp([a]) &= e^a \end{aligned} \quad (2.24)$$

Are there 3-dimensional types of complex numbers (division algebras, to be posh) that derive from the finite group $C_3$? Yes, there are four of them. There are division algebras that derive from every one of the infinite number of finite groups. All we have to do is write the finite group as a set of permutation matrices of the same size as the order of the finite group. Such a set of permutation matrices will always add to be a matrix with a single 1 as every element. We convert the 1's into real variables, add the matrices and take the exponential of the matrix (easier said than done for higher order finite groups). We will do exactly this when we derive the quaternions.

Summary:
Numbers are division algebras. Any set of mathematical objects which satisfies the division algebra axioms is a type of numbers. Multiplicative commutativity is not a division

algebra axiom, and there are many division algebras which are multiplicatively non-commutative.

All division algebras derive from a finite group, and all finite groups hold division algebras within them.

Algebraic extensions and monic minimal polynomials are, in your author's opinion, best consigned to a book on the history of how humans so easily mislead themselves. A fool once convinced is easily unconvinced; a wise person once convinced is stuck with it.

# Chapter 3

# Derivation of the Quaternions

We begin with the finite group $C_2 \times C_2$. This is a commutative finite group, and so it is remarkable that this finite group contains non-commutative division algebras such as the quaternions. It seems, there is no proof of this and so it might not be true, that non-commutative finite groups always give rise to non-commutative division algebras (non-commutative types of number) and that commutative groups always give rise to commutative division algebras (commutative types of number) with the exception of the commutative finite groups of the form $C_2 \times C_2 \times ...$ which give rise to both commutative division algebras and to non-commutative division algebras. These exceptions do not include cross products of any finite group other than $C_2$. The commutative finite groups $C_3 \times C_3$ and $C_4 \times C_4$ give rise to only commutative division algebras.

In this regard, the $C_2 \times C_2 \times ...$ finite groups are extremely exceptional. We remind the reader that there is not yet a proof of this conjecture.

The finite group $C_2 \times C_2$ is the permutation matrices:

$$\begin{bmatrix} 1 & 0 & 0 & 0 \\ 0 & 1 & 0 & 0 \\ 0 & 0 & 1 & 0 \\ 0 & 0 & 0 & 1 \end{bmatrix} \quad \begin{bmatrix} 0 & 1 & 0 & 0 \\ 1 & 0 & 0 & 0 \\ 0 & 0 & 0 & 1 \\ 0 & 0 & 1 & 0 \end{bmatrix} \quad (3.1)$$

$$\begin{bmatrix} 0 & 0 & 1 & 0 \\ 0 & 0 & 0 & 1 \\ 1 & 0 & 0 & 0 \\ 0 & 1 & 0 & 0 \end{bmatrix} \quad \begin{bmatrix} 0 & 0 & 0 & 1 \\ 0 & 0 & 1 & 0 \\ 0 & 1 & 0 & 0 \\ 1 & 0 & 0 & 0 \end{bmatrix} \quad (3.2)$$

Notice that these matrices are symmetric about the leading diagonal which runs from the top left-hand corner to the bottom right-hand corner of the matrix. Symmetric matrices have real eigenvalues and an orthogonal set of eigenvectors just like hermitian matrices. When we form the division algebras within the finite group $C_2 \times C_2$, we will 'scatter' a few minus signs about these matrices, but we will never produce a matrix with variables which are not either symmetric or anti-symmetric about the leading diagonal.

We take the four matrices (3.1) & (3.2), and we introduce real variables and we add them to produce the basic $C_2 \times C_2$ algebraic matrix form:

$$C_2 \times C_2 \rightarrow \begin{bmatrix} a & b & c & d \\ b & a & d & c \\ c & d & a & b \\ d & c & b & a \end{bmatrix} \quad (3.3)$$

*Derivation of the Quaternions*

In due process, we will take the exponential of this matrix, but, for now, we will use it in the above Cartesian form. The exponential of this matrix, (3.3), is a commutative division algebra, but we seek all the division algebras which derive from the finite group $C_2 \times C_2$. We want to know how to 'scatter' the minus signs in this matrix to produce the other division algebras that derive from the finite group $C_2 \times C_2$.

We begin by multiplying each element of the basic $C_2 \times C_2$ algebraic matrix form by a real variable[8] which we denote by $P_{\text{row,column}} \neq 0$. These real variables can take any real value other than zero, but, for our purposes, they will be either plus unity or minus unity:

$$\begin{bmatrix} P_{1,1}a & P_{1,2}b & P_{1,3}c & P_{1,4}d \\ P_{2,1}b & P_{2,2}a & P_{2,3}d & P_{2,4}c \\ P_{3,1}c & P_{3,2}d & P_{3,3}a & P_{3,4}b \\ P_{4,1}d & P_{4,2}c & P_{4,3}b & P_{4,4}a \end{bmatrix} \qquad (3.4)$$

These $P_{\text{row,column}}$ parameters will eventually be set to either +1 or −1 to form the division algebras.

We soon realise that the only effect of the parameters in the top row is to scale each variable, and so we can set these to +1. Further, every division algebra has to have a multiplicative identity, and so all the parameters on the

---

[8] Any mathematical object other than a real number, say a function, would violate the division algebra axioms in one way or another. A consequence of this is that the 'complex space' cannot have curvature. Division algebra spaces are held rigidly flat by the algebraic structure.

*Quaternions*

leading diagonal must be the same. Since we have set the leading diagonal parameter on the top row to +1, we will have to set the other leading diagonal parameters to +1. (We could have set them all to −1 if we had wanted.) We now have:

$$\begin{bmatrix} a & b & c & d \\ P_{2,1}b & a & P_{2,3}d & P_{2,4}c \\ P_{3,1}c & P_{3,2}d & a & P_{3,4}b \\ P_{4,1}d & P_{4,2}c & P_{4,3}b & a \end{bmatrix} \quad (3.5)$$

We form a copy of this matrix, but we change the variables. We will then form the product of these two matrices and insist upon multiplicative closure of form − a duck mated with a duck must produce a duck. We have:

$$\begin{bmatrix} a & b & c & d \\ P_{2,1}b & a & P_{2,3}d & P_{2,4}c \\ P_{3,1}c & P_{3,2}d & a & P_{3,4}b \\ P_{4,1}d & P_{4,2}c & P_{4,3}b & a \end{bmatrix} \begin{bmatrix} e & f & g & h \\ P_{2,1}f & e & P_{2,3}h & P_{2,4}g \\ P_{3,1}g & P_{3,2}h & e & P_{3,4}f \\ P_{4,1}h & P_{4,2}g & P_{4,3}f & e \end{bmatrix}$$

(3.6)

Although these matrices are not commutative, since we are seeking multiplicative closure, the order of multiplication does not matter for our purposes.

The leading diagonal elements of the product matrix are:

*Derivation of the Quaternions*

$$\begin{aligned}
&ae + P_{2,1}bf + P_{3,1}cg + P_{4,1}dh \\
&ae + P_{2,1}bf + P_{2,4}P_{4,2}cg + P_{2,3}P_{3,2}dh \\
&ae + P_{3,4}P_{4,3}bf + P_{3,1}cg + P_{2,3}P_{3,2}dh \\
&ae + P_{3,4}P_{4,3}bf + P_{2,4}P_{4,2}cg + P_{4,1}dh
\end{aligned} \quad (3.7)$$

These must all be equal if we are to have multiplicative closure. We thus must have:

$$P_{4,3} = \frac{P_{2,1}}{P_{3,4}}, \qquad P_{4,2} = \frac{P_{3,1}}{P_{2,4}}, \qquad P_{4,1} = P_{2,3}P_{3,2} \quad (3.8)$$

Notice that these equations are all linear; they have only one solution – this is important.

We could have chosen to eliminate other parameters instead of $P_{4,1}$, $P_{4,2}$, & $P_{4,3}$, but the essence of the result is the same. Actually, everything about the result is the same other than the appearance, and it is easy to move from one appearance to a different appearance by substituting for a particular parameter.

Substituting (3.8) into (3.5) gives:

$$\begin{bmatrix}
a & b & c & d \\
P_{2,1}b & a & P_{2,3}d & P_{2,4}c \\
P_{3,1}c & P_{3,2}d & a & P_{3,4}b \\
P_{2,3}P_{3,2}d & \dfrac{P_{3,1}}{P_{2,4}}c & \dfrac{P_{2,1}}{P_{3,4}}b & a
\end{bmatrix} \quad (3.9)$$

We form the product of two such matrices:

$$\begin{bmatrix} a & b & c & d \\ P_{2,1}b & a & P_{2,3}d & P_{2,4}c \\ P_{3,1}c & P_{3,2}d & a & P_{3,4}b \\ P_{2,3}P_{3,2}d & \dfrac{P_{3,1}}{P_{2,4}}c & \dfrac{P_{2,1}}{P_{3,4}}b & a \end{bmatrix} \begin{bmatrix} e & f & g & h \\ P_{2,1}f & e & P_{2,3}h & P_{2,4}g \\ P_{3,1}g & P_{3,2}h & e & P_{3,4}f \\ P_{2,3}P_{3,2}h & \dfrac{P_{3,1}}{P_{2,4}}g & \dfrac{P_{2,1}}{P_{3,4}}f & e \end{bmatrix}$$

(3.10)

To have multiplicative closure, the leftmost element of row two, $Z_{2,1}$, of the product matrix must be $P_{2,1}$ multiplied by the second leftmost element of the top row, $Z_{1,2}$. We have:

$$Z_{1,2} = af + be + P_{3,2}ch + \frac{P_{3,1}}{P_{2,4}}dg$$
$$P_{2,1}Z_{1,2} = P_{2,1}af + P_{2,1}be + P_{2,1}P_{3,2}ch + \frac{P_{2,1}P_{3,1}}{P_{2,4}}dg$$

(3.11)

We require the element $Z_{2,1}$ to be equal to the lower of the above equations, (3.11). This is:

$$P_{2,1}af + P_{2,1}be + P_{2,3}P_{2,4}P_{3,2}ch + P_{2,3}P_{3,1}dg =$$
$$P_{2,1}af + P_{2,1}be + P_{2,1}P_{3,2}ch + \frac{P_{2,1}P_{3,1}}{P_{2,4}}dg$$

(3.12)

From this, we have:

$$P_{2,4} = \frac{P_{2,1}}{P_{2,3}}$$

(3.13)

Similarly, comparing the element $Z_{3,1}$ gives:

*Derivation of the Quaternions*

$$P_{3,4} = \frac{P_{3,1}}{P_{3,2}} \qquad (3.14)$$

We have now reached a point where no matter which elements of the product matrix we compare, we can find no more linear equations that will eliminate another parameter. We have:

$$\begin{bmatrix} a & b & c & d \\ P_{2,1}b & a & P_{2,3}d & \dfrac{P_{2,1}}{P_{2,3}}c \\ P_{3,1}c & P_{3,2}d & a & \dfrac{P_{3,1}}{P_{3,2}}b \\ P_{2,3}P_{3,2}d & \dfrac{P_{2,3}P_{3,1}}{P_{2,1}}c & \dfrac{P_{2,1}P_{3,2}}{P_{3,1}}b & a \end{bmatrix} \qquad (3.15)$$

To eliminate another parameter, we are compelled to accept the quadratic equation:

$$P_{3,2}^{\;2} = \frac{P_{2,3}^{\;2} P_{3,1}^{\;2}}{P_{2,1}^{\;2}} \qquad (3.16)$$

This leads to two multiplicatively closed matrix forms. Taking the positive root gives:

$$\begin{bmatrix} a & b & c & d \\ P_{2,1}b & a & P_{2,3}d & \dfrac{P_{2,1}}{P_{2,3}}c \\ P_{3,1}c & \dfrac{P_{2,3}P_{3,1}}{P_{2,1}}d & a & \dfrac{P_{2,1}}{P_{2,3}}b \\ \dfrac{P_{2,3}^2 P_{3,1}}{P_{2,1}}d & \dfrac{P_{2,3}P_{3,1}}{P_{2,1}}c & P_{2,3}b & a \end{bmatrix} \quad (3.17)$$

And taking the negative root gives:

$$\begin{bmatrix} a & b & c & d \\ P_{2,1}b & a & P_{2,3}d & \dfrac{P_{2,1}}{P_{2,3}}c \\ P_{3,1}c & -\dfrac{P_{2,3}P_{3,1}}{P_{2,1}}d & a & -\dfrac{P_{2,1}}{P_{2,3}}b \\ -\dfrac{P_{2,3}^2 P_{3,1}}{P_{2,1}}d & \dfrac{P_{2,3}P_{3,1}}{P_{2,1}}c & -P_{2,3}b & a \end{bmatrix} \quad (3.18)$$

The first of these two forms, (3.17), is a multiplicatively commutative matrix form. The second of these, (3.18), is a multiplicatively non-commutative matrix form. We have found non-commutative division algebras within a commutative group – that is remarkable.

If we had chosen the finite group $C_3$ or the finite group $C_4$ or the finite group $C_3 \times C_3$ or the finite group $C_4 \times C_4$ instead of choosing $C_2 \times C_2$, we would have been able to eliminate the number of parameters necessary for

## Derivation of the Quaternions

multiplicative closure using only linear equations and we would have only commutative division algebras. It seems, and there is no proof of this conjecture, that other than the finite group $C_2 \times C_2$ and groups of which $C_2 \times C_2$ is a subgroup such as the $C_2 \times C_2 \times C_2 \times ...$ groups, the appearance of a non-linear elimination equation never occurs.

Although it is not the purpose of this book to discuss classical physics, your author must point out that the classical physics of general relativity, Riemann geometry, an expanding universe, and classical electromagnetism derives from six of the eight division algebras within the non-commutative matrix form (3.18)[9]. None of this classical physics would exist if it were not for the minus sign in (3.16).

The reader might have noticed that we could have chosen to eliminate parameters in a different way and have been left with a different set of parameters. However we choose to eliminate parameters leads to a result which is effectively the same as the result above. For example, if we substitute into (3.17) (or (3.18)) we get:

---

[9] See: Dennis Morris: Upon General Relativity.

*Quaternions*

$$P_{2,3} = \frac{P_{2,1}}{P_{2,4}}$$

$$\begin{bmatrix} a & b & c & d \\ P_{2,1}b & a & P_{2,3}d & P_{2,4}c \\ P_{3,1}c & \frac{P_{3,1}}{P_{2,4}}d & a & P_{2,4}b \\ \frac{P_{2,1}P_{3,1}}{P_{2,4}^2}d & \frac{P_{3,1}}{P_{2,4}}c & \frac{P_{2,1}}{P_{2,4}}b & a \end{bmatrix} \quad (3.19)$$

Forming the division algebras:

We have no interest in the eight commutative algebras that we get from (3.17), and so we begin with (3.18). By setting the parameters to the eight permutations of $\pm 1$, we get the eight division algebras.

First, we present the six $A_3$ algebras:

$$P_{2,1} = +1, \quad P_{2,3} = +1, \quad P_{3,1} = +1,$$

$$SSA^*_{Anti} = \exp\left(\begin{bmatrix} a & b & c & d \\ b & a & d & c \\ c & -d & a & -b \\ -d & c & -b & a \end{bmatrix}\right) \quad (3.20)$$

*Derivation of the Quaternions*

$$P_{2,1} = +1, \quad P_{2,3} = -1, \quad P_{3,1} = +1,$$

$$SSA^* = \exp\left(\begin{bmatrix} a & b & c & d \\ b & a & -d & -c \\ c & d & a & b \\ -d & -c & b & a \end{bmatrix}\right) \quad (3.21)$$

$$P_{2,1} = +1, \quad P_{2,3} = +1, \quad P_{3,1} = -1,$$

$$SAS = \exp\left(\begin{bmatrix} a & b & c & d \\ b & a & d & c \\ -c & d & a & -b \\ d & -c & -b & a \end{bmatrix}\right) \quad (3.22)$$

$$P_{2,1} = +1, \quad P_{2,3} = -1, \quad P_{3,1} = -1,$$

$$SAS_{Anti} = \exp\left(\begin{bmatrix} a & b & c & d \\ b & a & -d & -c \\ -c & -d & a & b \\ d & c & b & a \end{bmatrix}\right) \quad (3.23)$$

$$P_{2,1} = -1, \quad P_{2,3} = -1, \quad P_{3,1} = +1,$$

$$ASS = \exp\left(\begin{bmatrix} a & b & c & d \\ -b & a & -d & c \\ c & -d & a & -b \\ d & c & b & a \end{bmatrix}\right) \quad (3.24)$$

$$P_{2,1} = -1, \quad P_{2,3} = +1, \quad P_{3,1} = +1,$$

$$ASS_{Anti} = \exp\left(\begin{bmatrix} a & b & c & d \\ -b & a & d & -c \\ c & d & a & b \\ d & -c & -b & a \end{bmatrix}\right) \quad (3.25)$$

Now we present the two quaternion algebras:

$$P_{2,1} = -1, \quad P_{2,3} = -1, \quad P_{3,1} = -1,$$

$$\mathbb{H} = \begin{bmatrix} a & b & c & d \\ -b & a & -d & c \\ -c & d & a & -b \\ -d & -c & b & a \end{bmatrix} \quad (3.26)$$

$$P_{2,1} = -1, \quad P_{2,3} = +1, \quad P_{3,1} = -1,$$

$$\mathbb{H}_{Anti} = \begin{bmatrix} a & b & c & d \\ -b & a & d & -c \\ -c & -d & a & b \\ -d & c & -b & a \end{bmatrix} \quad (3.27)$$

The particular permutations reflect which parameters we chose to eliminate.

The determinants, norms, of the $A_3$ algebras are of the form:

$$n^2 = a^2 + b^2 - c^2 - d^2 \quad (3.28)$$

*Derivation of the Quaternions*

Since these determinants can be zero, the $A_3$ matrices become algebras only when we take the exponential to be rid of the singular matrices.

The determinants, norms, of the quaternion algebras are:

$$n^2 = a^2 + b^2 + c^2 + d^2 \qquad (3.29)$$

And so the quaternion matrices can never be singular, and so there is no need to take the exponential to ensure the existence of multiplicative inverses.

We have now derived the two quaternion algebras, (3.26) & (3.27).

The distance function of quaternion (anti-quaternion) space is the norm of the quaternions (anti-quaternions) which is the determinant given by (3.29)

With a little consideration, the reader might realise that we could have obtained the two quaternion algebras by simply combining three anti-symmetric variables in every possible way that is multiplicatively closed. The same thought applies to the six $A_3$ algebras except that we use one anti-symmetric variable and two symmetric variables. The two types of commutative division algebras, the $A_1$ & $A_2$ algebras that derive from the group $C_2 \times C_2$ and which we discarded above are formed from (2 off) three symmetric variables and (six off) two anti-symmetric variables and one symmetric variable.

## Quaternions

Emergent expectation distance functions:

There are an infinite number of different types of division algebras within the infinite number of finite groups. Many of these division algebras are algebraically isomorphic to each other. Above, we have seen two quaternion algebras and six $A_3$ algebras emerge from the $C_2 \times C_2$ finite group. That is eight separate algebras, but they are of only two algebraically distinct types.

Every division algebra has rotation within it – the polar form of the algebra. Each type of rotation preserves the distance function (determinant) of the algebra – that's what rotation does; it holds the distance from the origin invariant.

We form the emergent expectation distance function of a particular division algebra by adding the distance functions of all the isomorphic copies of that particular division algebra that derive from a finite group. For example, if we add the distance functions of the six $A_3$ algebras, we get:

$$\text{Sum} \begin{cases} d^2 = t^2 + x^2 - y^2 - z^2 \\ d^2 = t^2 + x^2 - y^2 - z^2 \\ d^2 = t^2 - x^2 + y^2 - z^2 \\ d^2 = t^2 - x^2 + y^2 - z^2 \\ d^2 = t^2 - x^2 - y^2 + z^2 \\ d^2 = t^2 - x^2 - y^2 + z^2 \end{cases} \quad (3.30)$$

$$6d^2 = 2\left(3t^2 - x^2 - y^2 - z^2\right)$$

The 2 and the 6 is just a scaling factor that is meaningless, and the 3 is just the units chosen to measure time or distance.

*Derivation of the Quaternions*

We have the distance function of our 4-dimensional space-time.

This emergent expectation distance function can support both types of 2-dimensional rotation because these types of 2-dimensional rotation preserve the 2-dimensional distance functions $d^2 = t^2 - z^2$ and $d^2 = x^2 + y^2$ and these distance functions are (three of each type) sub-functions of the 4-dimensional distance function[10].

The sum of the two quaternion distance functions is:

$$Sum \begin{cases} d^2 = t^2 + x^2 + y^2 + z^2 \\ d^2 = t^2 + x^2 + y^2 + z^2 \end{cases} \quad (3.31)$$

$$2d^2 = 2\left(t^2 + x^2 + y^2 + z^2\right)$$

This emergent expectation distance function can support 2-dimensional Euclidean rotations (six off) and it can also support quaternion rotation, but it cannot support rotation in 2-dimensional space-time (no minus signs in the distance function).

We have an infinite number of distance functions within the division algebras which derive from the infinite number of finite groups. We therefore have an infinite number of emergent expectation distance functions. Remarkably, it seems, and there is not yet a proof of this, that only the $A_3$ emergent expectation distance function and the quaternion emergent expectation distance function are able to support any of the other division algebra rotations; in other words, it

---

[10] This is why we have 2-dimensional rotations of two types in our one type of 4-dimensional space-time.

seems that, of an infinite number of emergent expectation spaces, there are only two which have rotation within them, but remember this has not yet been proven. None-the-less, you were wise to buy this book.

Square roots of real numbers:
The square root of the expression:

$$\begin{bmatrix} x^2+1 & 0 \\ 0 & x^2+1 \end{bmatrix} \quad (3.32)$$

is the two complex numbers:

$$\begin{bmatrix} x & -1 \\ 1 & x \end{bmatrix} \begin{bmatrix} x & 1 \\ -1 & x \end{bmatrix} = \begin{bmatrix} x^2+1 & 0 \\ 0 & x^2+1 \end{bmatrix} \quad (3.33)$$

In general, the square root of a real number, if not a real number, is a type of complex number multiplied its conjugate. Thinking of the complex plane, $\mathbb{C}$, we see a complex number and its conjugate as being points in the complex plane which are equidistant from the origin and are at equal angles from the real axis. Multiplying a complex number by its conjugate always gives a real number. Thus, the square root of a real number, real expression, if a complex number, is of the form $\mathbb{C}^*\mathbb{C}$ rather than of the form $\mathbb{C}\mathbb{C}$.

Similarly with quaternions, the square root of a real number, real expression, is a quaternion multiplied by its conjugate. An example is:

*Derivation of the Quaternions*

$$\begin{bmatrix} m & -p_x & -p_y & -p_z \\ p_x & m & p_z & -p_y \\ p_y & -p_z & m & p_x \\ p_z & p_y & -p_x & m \end{bmatrix} \begin{bmatrix} m & p_x & p_y & p_z \\ -p_x & m & -p_z & p_y \\ -p_y & p_z & m & -p_x \\ -p_z & -p_y & p_x & m \end{bmatrix}$$

$$= \begin{bmatrix} p_x^2 + p_y^2 + p_z^2 + m^2 & 0 & 0 & 0 \\ 0 & \sim & 0 & 0 \\ 0 & 0 & \sim & 0 \\ 0 & 0 & 0 & p_x^2 + p_y^2 + p_z^2 + m^2 \end{bmatrix}$$

(3.34)

The expression in the product matrix is the RHS of the relativistic energy momentum relation:

$$E^2 = p_x^2 + p_y^2 + p_z^2 + m^2 \qquad (3.35)$$

That a quaternion is the square root of the RHS of this expression, (3.35), is intimately connected to the Dirac equation of quantum electro-dynamics and the quaternion like nature gamma matrices within the Dirac equation, but that is too far from our present concerns to discuss in detail.

*Quaternions*

## Chapter 4

# Conventional Notation

The quaternions, as discovered by William Hamilton, are conventionally written as:

$$\mathbb{H} = a + \hat{i}b + \hat{j}c + \hat{k}d$$
$$\hat{i}^2 = -1, \quad \hat{j}^2 = -1, \quad \hat{k}^2 = -1$$
$$\hat{i}\hat{j} = \hat{k}, \quad \hat{j}\hat{k} = \hat{i}, \quad \hat{k}\hat{i} = \hat{j} \quad (4.1)$$
$$\hat{j}\hat{i} = -\hat{k}, \quad \hat{k}\hat{j} = -\hat{i}, \quad \hat{i}\hat{k} = -\hat{j},$$

The matrix notation above, (3.26), uses the same $\{a,b,c,d\}$ variables but has the advantage that all the relations between the unit imaginary variables are automatically included by the nature of matrix multiplication[11].

The anti-quaternions, were not discovered until recently. The anti-quaternions are algebraically isomorphic to the quaternions. They are the same algebra written in a different basis. However, the anti-quaternions must not be ignored; they play an important part in physics; the six $A_3$ algebras are all algebraically isomorphic, but we need all six to derive classical physics. The commutation relations of the anti-

---

[11] This is an example of how important it can be to get the correct notation. Choosing the correct notation goes hand-in-hand with properly understanding with what one is dealing.

quaternions are the reverse of the commutation relations of the quaternions. We have:

$$\mathbb{H}_{Anti} = a + \hat{i}b + \hat{j}c + \hat{k}d$$
$$\hat{i}^2 = -1, \quad \hat{j}^2 = -1, \quad \hat{k}^2 = -1$$
$$\hat{i}\hat{j} = -\hat{k}, \quad \hat{j}\hat{k} = -\hat{i}, \quad \hat{k}\hat{i} = -\hat{j}$$
$$\hat{j}\hat{i} = \hat{k}, \quad \hat{k}\hat{j} = \hat{i}, \quad \hat{i}\hat{k} = \hat{j},$$
(4.2)

The matrix notation above, (3.27), uses the same $\{a, b, c, d\}$ variables but has the advantage that all the relations are automatically included by the nature of matrix multiplication.

## Chapter 5

# Quaternion Rotation

By taking the exponential of the imaginary variable of the Euclidean complex numbers, we get the rotation matrix of the complex plane, $\mathbb{C}$:

$$\exp\left(\begin{bmatrix} 0 & b \\ -b & 0 \end{bmatrix}\right) = \begin{bmatrix} \cos b & \sin b \\ -\sin b & \cos b \end{bmatrix} \quad (5.1)$$

The radial variable is from the real variable:

$$\exp\left(\begin{bmatrix} a & 0 \\ 0 & a \end{bmatrix}\right) = \begin{bmatrix} e^a & 0 \\ 0 & e^a \end{bmatrix} \quad (5.2)$$

The Euclidean complex plane is 2-dimensional; there is no third dimension sticking out of the plane at right-angles to it. The above rotation matrix is a 2-dimensional rotation within a 2-dimensional space; there is no third dimension to be the axis of rotation. This is not rotation about an axis.

"But I thought all rotation was rotation about an axis", we hear the reader cry. Well, you thought wrongly. Rotation in every division algebra is of the same nature as it is within the complex plane in that it is never rotation about an axis. We get rotation about an axis in only emergent expectation spaces and then only if the expectation distance function of that emergent expectation space can support lesser dimensional rotations.

## Quaternion Rotation

Mathematically, we can show that rotation in the complex plane, $\mathbb{C}$, is not about an axis by taking the eigenvalues (or eigenvectors) of the rotation matrix, (5.1). The eigenvalues of (5.1) are $e^{ib}$ & $e^{-ib}$. Clearly, these eigenvalues are not independent of the rotation angle, $b$, and so we have no direction in this space (axis) which is unaffected by the rotation; we have no axis of rotation.

Of course, to be a little tautological, the rotation matrices which are the angular part of the polar form of any division algebra are, as with the Euclidean complex numbers, a n-dimensional rotation in a n-dimensional space and thus are not rotations about an axis. The only rotations about an axis are the 2-dimensional rotations in our 4-dimensional space-time and the 2-dimensional rotations in the quaternion emergent expectation space. We have rotation about an axis (two axes actually) within our 4-dimensional space-time because the distance function of our 4-dimensional space-time supports rotations (2-dimensional) of a lesser dimension than itself and thereby leaves other dimensions unaffected by the rotation.

Taking the eigenvectors of any of the six rotation matrices that specify 2-dimensional rotation in our 4-dimensional space-time will deliver two eigenvectors which are independent of the rotation angle. For example, we have:

$$\begin{bmatrix} \cosh \chi_1 & \sinh \chi_1 & 0 & 0 \\ \sinh \chi_1 & \cosh \chi_1 & 0 & 0 \\ 0 & 0 & 1 & 0 \\ 0 & 0 & 0 & 1 \end{bmatrix} \sim \begin{bmatrix} e^{\chi_1} \\ 0 \\ 0 \\ 0 \end{bmatrix}, \begin{bmatrix} 0 \\ e^{\chi_1} \\ 0 \\ 0 \end{bmatrix}, \begin{bmatrix} 0 \\ 0 \\ 1 \\ 0 \end{bmatrix}, \begin{bmatrix} 0 \\ 0 \\ 0 \\ 1 \end{bmatrix}$$

(5.3)

## Connections between trig. functions:

There is a thought which we should keep in our minds. We find the trigonometric functions $\sin(\ )\ \&\cos(\ )$ used all over the place in mathematics. It is sobering to realise that these functions exist in only the Euclidean complex plane; something very much like them exists in the quaternion space. These functions might not exist outside of the Euclidean complex plane, but they often appear as a part of other functions.

An example of such an appearance is within the 3-dimensional trigonometric functions which derive from the finite group $C_3$. There is no 2-dimensional rotation of any kind in the 3-dimensional $C_3$ spaces because $C_2$ is not a sub-group of $C_3$, and so the $\sin(\ )\ \&\cos(\ )$ functions have no place in these spaces. In one sense, the $\sin(\ )\ \&\cos(\ )$ functions appear in the 3-dimensional trigonometric functions partly by the 'accident' of the limitations our list of functions, but there is a deeper reason for their appearance here. The exponential function underlies all trigonometric functions of all finite groups because it is by taking the exponential of the algebraic matrix form of the group that we get the rotation matrices of that group. The 3-dimensional trigonometric functions are connected to the 2-dimensional trigonometric functions, and to the trigonometric functions of all other dimensions, via the exponential function. This is why we find the $\sin(\ )\ \&\cos(\ )$ functions within spaces where they have no place. If we were to change our list of functions to include individual names for the 3-dimensional trigonometric functions, say $\{v_A, v_B, v_C\}$, but to exclude the sin( ) & cos( ) functions, then the trigonometric functions of the Euclidean complex plane would be expressed as functions of $\{v_A, v_B, v_C\}$.

## The polar form of the quaternions:

The radial variable in the polar form of the quaternions derives from taking the exponential of the real variable. Since the real variable commutes with each imaginary variable, separating it from the imaginary variables presents no problem when taking the matrix exponential. In the quaternion case, we have:

$$\exp\left(\begin{bmatrix} 0 & b & c & d \\ -b & 0 & -d & c \\ -c & d & 0 & -b \\ -d & -c & b & 0 \end{bmatrix}\right) \qquad (5.4)$$

This is:

$$\mathbb{H}_{Rot} = \begin{bmatrix} \cos\lambda & \frac{b}{\lambda}\sin\lambda & \frac{c}{\lambda}\sin\lambda & \frac{d}{\lambda}\sin\lambda \\ -\frac{b}{\lambda}\sin\lambda & \cos\lambda & -\frac{d}{\lambda}\sin\lambda & \frac{c}{\lambda}\sin\lambda \\ -\frac{c}{\lambda}\sin\lambda & \frac{d}{\lambda}\sin\lambda & \cos\lambda & -\frac{b}{\lambda}\sin\lambda \\ -\frac{d}{\lambda}\sin\lambda & -\frac{c}{\lambda}\sin\lambda & \frac{b}{\lambda}\sin\lambda & \cos\lambda \end{bmatrix}$$

(5.5)

*Quaternions*

$$\lambda = \sqrt{b^2 + c^2 + d^2} \qquad (5.6)$$

This is the rotation matrix of quaternion space. Again, this is not rotation about an axis. More extraordinary than that, this is a double cover rotation. We will look at double cover rotation in more detail shortly.

No rotation axis:

The eigenvalues of $\mathbb{H}_{Rot}$ are:

$$\begin{array}{ll} e^{\sqrt{-1}\sqrt{b^2+c^2+d^2}} & \textit{Twice} \\ e^{-\sqrt{-1}\sqrt{b^2+c^2+d^2}} & \textit{Twice} \end{array} \qquad (5.7)$$

Clearly, these vary with the angle, and so this is not rotation about an axis; we knew that anyway.

We can adjust the co-ordinate system so that this rotation happens in, say, the 2-dimensional plane formed from the real axis and the imaginary $b$ axis. We then have:

$$\mathbb{H}_{Rot} = \begin{bmatrix} \cos\lambda & \sqrt{1}\sin\lambda & 0 & 0 \\ -\sqrt{1}\sin\lambda & \cos\lambda & 0 & 0 \\ 0 & 0 & \cos\lambda & -\sqrt{1}\sin\lambda \\ 0 & 0 & \sqrt{1}\sin\lambda & \cos\lambda \end{bmatrix}$$

$$\lambda = \sqrt{b^2}$$
(5.8)

Although this is rotation in a 2-dimensional plane within the 4-dimensional quaternion space, it is not 2-dimensional

## Quaternion Rotation

rotation about an axis (or two). The eigenvalues of this rotation matrix are just the eigenvalues given above, (5.7), with $c = d = 0$, and so this is not a 2-dimensional rotation about an axis in 4-dimensional space. This is a 4-dimensional rotation that just happens to be in a particular co-ordinate system.

Now, within our 4-dimensional space-time, we have only two types of rotation because the distance function of our 4-dimensional space-time admits only those two types of rotation. The rotations within our 4-dimensional space-time are both 2-dimensional rotations from the $C_2$ group; these are the Lorentz boost (3 off) and the 2-dimensional Euclidean rotation (3 off). How might we beings in 4-dimensional space-time perceive a 4-dimensional quaternion rotation? The only rotations we can see are the 2-dimensional rotations, and, of these, the 2-dimensional Euclidean rotation seems to be the only type of rotation that would fit. By this we mean that form of the distance function of the quaternions, $d^2 = t^2 + x^2 + y^2 + z^2$, reduces to the form, $d^2 = y^2 + z^2$, of the 2-dimensional Euclidean space when two of the variables are zero.

### Double cover rotation:

Looking at (5.8), we notice that we have a clockwise rotation in the top left-hand corner and an anti-clockwise rotation in the bottom right-hand corner. Your author opines that these two sub-rotations are the origin of the double cover rotation associated with the quaternions as $SU(2)$. Your author previously opined that it was the square root in the trigonometric functions together with $\sin(-b) = -\sin(b)$ that led to double cover, but your author now believes this is incorrect and that we should consistently take the same sign under the square roots[12]. Let us rotate through the angle, say, $b = \dfrac{\pi}{4} \equiv 45^0$. Because, in our 4-dimensional space-time, we see only 2-dimensional rotations, we see two 2-dimensional rotations in this quaternion rotation. We get 2-dimensional rotation in both the clockwise direction and the anti-clockwise direction at the same time.

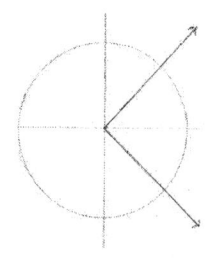

We have twice as much rotation as is morally proper. Perhaps we can salve our shock by appreciating the symmetry of quaternion rotations as compared to the 'lop-sided' rotation to which we are accustomed.

Another way to view this 'double rotation' is that we can get a clockwise rotation of say $\dfrac{\pi}{4} \equiv 45^0$ by putting in two angles, $b = \pm\dfrac{\pi}{4} \equiv \pm 45^0$. Two different angles give the same rotation.

---

[12] These are opinions not facts. You are welcome to form your own opinion.

Of course, all of the above applies if we re-set the co-ordinate system so the rotation happens in the plane formed with the real axis and a different imaginary variable. This phenomenon of double rotation is called double cover, but, looking at the literature, you might not realise that this is the double cover to which the literature refers. Note that the rotation is simply connected because we are within a division algebra.

Violation of parity perhaps:
Looking at the quaternion rotation matrix, (5.5), the reader will see that rotation with only the $d$ variable is both clockwise and anti-clockwise as it is with only the $b$ variable but rotation with the $c$ variable is clockwise twice. Looking at the distribution of minus sings in the anti-quaternion, (3.27), we see that we will get rotation in both clockwise and anti-clockwise directions with only the $c$ variable and we get rotation in the clockwise direction twice with the other two variables. In this regard, there is an imbalance between the quaternions and the anti-quaternions. Is this the parity violation we associate with the weak force? We do not know.

Quaternion rotation as a continuous group:
Within our 4-dimensional space-time, there are six possible pairings of variables into six 2-dimensional planes; three of these 2-dimensional planes are Euclidean, and three of these 2-dimensional planes are space-time planes. Within each 2-dimensional plane, there is a 2-dimensional rotation matrix (set into a $4 \times 4$ matrix with ten zeros and two ones); three

of these 2-dimensional rotation matrices are Euclidean rotations, and the other three 2-dimensional rotations are Lorentz boosts (space-time rotations) – six matrices altogether. Each 2-dimensional rotation matrix has a single parameter which is a single real number which we call the angle. That is six real numbers altogether. We can rotate in all six 2-dimensional planes within our 4-dimensional space-time because there are six continuous real variables we call angles.

In 4-dimensional quaternion space, there are again six pairs of variables which might form six 2-dimensional planes, but the quaternion rotation matrix holds only three real continuous variables. Therefore, in 4-dimensional quaternion space, we can rotate (4-dimensionally) 2-dimensionally in only three of the six 2-dimensional planes. The three planes in which we can rotate in quaternion space correspond to the three sub-algebras of the quaternions comprised of the identity element (the real part) and one of the imaginary variables. Quaternion space is not a space in which we can wave our arms around like we can in our 4-dimensional space-time; it is not a geometric space – we do not expect to observe it as a geometric space[13]. Yet the quaternion rotation matrix does define some form of continuous surface which is a continuous group.

The reader should try to picture quaternion rotation in 3-dimensional space by dispensing with one of the imaginary variables. In such a fantasy 3-dimensional quaternion space, we can rotate in two 2-dimensional planes corresponding to the pairs of variables which are the real variable and one of

---

[13] For much more about what a geometric space is, see: Dennis Morris & Sophie Lacson The Uniqueness of our Space-time.

the imaginary variables but we cannot rotate in, say the horizontal plane, formed from the two imaginary variables. Nor can we adjust the co-ordinate system by rotation in the, say, horizontal plane. We find that, by rotation, we are able to visit only four points on the horizontal 'circle' plane. These four points are two diametrically opposed pairs. This is electron spin quantitisation in a fantasy 3-dimensional space.

When we move back to the full 4-dimensional quaternion space, we have all six possible quantitised electron spins. That is why electron intrinsic spin is quantitised. Of course, the quaternions have the $SU(2)$ commutation relations, and they have the double cover rotation; both of these we associate with electrons. The continuous group which is the quaternion rotation matrix is equivalent to, but more clearly presented than, the Lie group $SU(2)$.

Restricted rotation in the quaternion space:
Suppose we try to rotate 4-dimensionally in quaternion space; that is we rotate with three non-zero variables. Can we do this?

Back to our fantasy 3-dimensional quaternion space. Let us rotate in both vertical planes at the same time. We find that after we have rotated through $90^0$ in both planes we have actually rotated through $45^0$ in the horizontal plane. But we cannot rotate in the horizontal plane. We see that even a very small rotation in the two vertical planes is a little rotation in the horizontal plane. This is Euler's theorem that any two 2-

dimensional rotations combine to form a single 2-dimensional rotation[14].

Returning to 4-dimensional quaternion space, we see that, because we have only three continuous variables (angles), any rotation in quaternion space must be a rotation with two of the three variables set equal to zero. We can have only 2-dimensional 4-dimensional rotations within quaternion space. These are still not rotations about an axis. Such rotations are in only one of the three planes corresponding to the three variables and never in a mixture of two planes. This is reminiscent of electron spin being directionally quantitised.

Non-commutative quaternion rotation:

Consider the rotation matrix of the 2-dimensional Euclidean complex plane, (5.1). Because the Euclidean complex numbers are commutative, a rotation matrix on the right will produce the same rotation as a rotation matrix on the left:

$$\begin{bmatrix} \cos\theta & \sin\theta \\ -\sin\theta & \cos\theta \end{bmatrix} \begin{bmatrix} x & y \\ -y & x \end{bmatrix} = \begin{bmatrix} x & y \\ -y & x \end{bmatrix} \begin{bmatrix} \cos\theta & \sin\theta \\ -\sin\theta & \cos\theta \end{bmatrix}$$
(5.9)

The quaternions are non-commutative. This means that a rotation matrix on the left will not give the same rotation as a rotation matrix on the right.

A quaternion is a position in 4-dimensional quaternion space just as a complex number is a position in the complex plane. A position is a vector, and so we can think of a quaternion

---

[14] Of course, Euler knew of only 2-dimensional rotations.

## Quaternion Rotation

as a vector in quaternion space. We will multiply the quaternion (vector):

$$Q = \begin{bmatrix} t & x & y & z \\ -x & t & -z & y \\ -y & z & t & -x \\ -z & -y & x & t \end{bmatrix} \quad (5.10)$$

by the rotation matrix (5.6). We get a quaternion, of course. We will present the top row of the product quaternion for ease. First, we multiply on the left:

$$\mathbb{H}_{Rot}Q_{[1,1]} = t\cos\lambda - \frac{1}{\lambda}\sin\lambda\left(bx + cy + dz\right)$$

$$\mathbb{H}_{Rot}Q_{[1,2]} = x\cos\lambda + \frac{1}{\lambda}\sin\lambda\left(bt + cz - dy\right)$$

$$\mathbb{H}_{Rot}Q_{[1,3]} = y\cos\lambda + \frac{1}{\lambda}\sin\lambda\left(-bz + ct + dx\right)$$

$$\mathbb{H}_{Rot}Q_{[1,4]} = z\cos\lambda + \frac{1}{\lambda}\sin\lambda\left(by - cx + dt\right)$$

(5.11)

Second, we multiply on the right:

$$Q\mathbb{H}_{Rot[1,1]} = t\cos\lambda - \frac{1}{\lambda}\sin\lambda\left(bx + cy + dz\right)$$

$$Q\mathbb{H}_{Rot[1,2]} = x\cos\lambda + \frac{1}{\lambda}\sin\lambda\left(bt - cz + dy\right)$$

$$Q\mathbb{H}_{Rot[1,3]} = y\cos\lambda + \frac{1}{\lambda}\sin\lambda\left(bz + ct - dx\right)$$

$$Q\mathbb{H}_{Rot[1,4]} = z\cos\lambda + \frac{1}{\lambda}\sin\lambda\left(-by + cx + dt\right)$$

(5.12)

*Quaternions*

We see that, because the quaternions are non-commutative, rotation on the left produces a different rotation from rotation on the right; but wait; we have forgotten that there is a square root sign in the trigonometric functions; will that help? No, it will not help. Looking at (5.11) & (5.12), whichever sign we choose for $\lambda = \sqrt{x^2 + y^2 + z^2}$ will not make them equal.

The real parts are certainly equal, but the imaginary parts are different, if only by a few signs. Nor will both clockwise and anti-clockwise rotation at the same time rescue the situation. We have to accept that a single rotation matrix through a given angle will rotate a vector in quaternion space to two different places depending on whether the quaternion rotation matrix is placed to the left of the quaternion vector or to the right of the quaternion vector.

We can reduce these rotations, (5.11) & (5.12), to the three 4-dimensional 2-dimensional rotations between the real axis and one imaginary axis. We get, (5.11):

$$\mathbb{H}_{Rot}Q_{[1,1]} = t\cos\sqrt{b^2} - \frac{x}{\sqrt{1}}\sin\sqrt{b^2}$$

$$\mathbb{H}_{Rot}Q_{[1,2]} = x\cos\sqrt{b^2} + \frac{t}{\sqrt{1}}\sin\sqrt{b^2}$$

$$\mathbb{H}_{Rot}Q_{[1,3]} = y\cos\sqrt{c^2} + \frac{t}{\sqrt{c^2}}\sin\sqrt{c^2} \qquad (5.13)$$

$$\mathbb{H}_{Rot}Q_{[1,4]} = z\cos\sqrt{d^2} + \frac{t}{\sqrt{1}}\sin\sqrt{d^2}$$

Except for the square root, taken separately, these are our familiar 2-dimensional rotations if we notice that we could

## Quaternion Rotation

have reduced the first of these differently to fit with the third or fourth instead of the second. We get the same results for (5.12) which is not surprising since 2-dimensional rotations commute.

Amazing things these quaternions!

Now, who says that rotation within a space should be done with only a single rotation matrix? People who are familiar with only the 2-dimensional commutative rotations say this. Perhaps we should try something different for non-commutative spaces.

### A non-commutative form of rotation:

There is much about non-commutative division algebras that is different from commutative division algebras; we will see shortly that differentiation is different for a start. Within our 4-dimensional space-time, we have experience of only commutative rotations (the 2-dimensional rotations). Perhaps our experience misleads us and the appropriate rotation to associate with non-commutative division algebras is non-commutative rotation.

Rotation is very closely connected to differentiation. The exponential function is such that its differential is equal to itself. We could see rotation matrices as just the exponential function in dimensions higher than one. We present an example:

$$\begin{bmatrix} \cosh \chi & \sinh \chi \\ \sinh \chi & \cosh \chi \end{bmatrix} \quad (5.14)$$

$$\cosh \chi + \sinh \chi = \exp \chi$$

*Quaternions*

Let us define a 2-dimensional Euclidean unit circle in the $\{x, y\}$ Euclidean plane as the locus of a point whose:

a) $x$ co-ordinate is the rate of change of the $y$ co-ordinate with respect to the angle subtended at the origin between the $x$ axis and the point

b) $y$ co-ordinate is the rate of change of the $x$ co-ordinate with respect to the angle subtended at the origin between the $x$ axis and the point

c) $x^2 + y^2 = 1$.

When we realise that the $x$ co-ordinate is just the cosine of the angle between the $x$ axis and the point and that the $y$ co-ordinate is just the sine of the angle between the $x$ axis and the point and that the differential of the sine is the cosine and that the differential of the cosine is the minus sine, we see that this is a valid definition of a unit circle in 2-dimensional Euclidean space.

Within the quaternions, we will soon see that we have to differentiate non-commutatively (next chapter). The non-commutative differentiation (see later) leads to two quaternion fields, the E-field and the B-field, formed from the left-differential, $d_L$, and the right-differential, $d_R$, as:

$$E = \frac{1}{2}(d_L + d_R)$$
$$B = \frac{1}{2}(d_L - d_R)$$
(5.15)

The direction of the B-field is arbitrary, and we could subtract the left-differential from the right-differential if we choose, but the above is conventional. We could treat rotation similarly to form a E-rotation and a B-rotation.

We will form the E-rotation as:

$$E^{Rot} = \frac{1}{2}\left(Rot_L + Rot_R\right)$$
$$B^{Rot} = \frac{1}{2}\left(Rot_L - Rot_R\right)$$
(5.16)

wherein $Rot_L$ is the product of multiplying a quaternion vector on the left by the rotation matrix and $Rot_R$ is the product of multiplying a quaternion vector on the right by the rotation matrix. Of course, the angle in the rotation matrix is the same in both cases. Using, (5.11) & (5.12) leads to the E-rotation:

$$E^{Rot}_{[1,1]} = \frac{1}{2}\left(\mathbb{H}_{Rot}\mathcal{Q}_{[1,1]} + \mathcal{Q}\mathbb{H}_{Rot[1,1]}\right)$$
$$= t\cos\lambda - \frac{1}{\lambda}\sin\lambda\left(bx + cy + dz\right)$$
(5.17)

$$E_{[1,2]}^{Rot} = \frac{1}{2}\left(\mathbb{H}_{Rot}\mathcal{Q}_{[1,2]} + \mathcal{Q}\mathbb{H}_{Rot[1,2]}\right)$$

$$= x\cos\lambda + \frac{bt}{\lambda}\sin\lambda$$

$$E_{[1,3]}^{Rot} = \frac{1}{2}\left(\mathbb{H}_{Rot}\mathcal{Q}_{[1,3]} + \mathcal{Q}\mathbb{H}_{Rot[1,3]}\right)$$

$$= y\cos\lambda + \frac{ct}{\lambda}\sin\lambda \quad (5.18)$$

$$E_{[1,4]}^{Rot} = \frac{1}{2}\left(\mathbb{H}_{Rot}\mathcal{Q}_{[1,4]} + \mathcal{Q}\mathbb{H}_{Rot[1,4]}\right)$$

$$= z\cos\lambda + \frac{dt}{\lambda}\sin\lambda$$

We compare this to (5.13) and see that the E-rotation is the condensation of the commutative 2-dimensional rotations.

The B-rotation is:

$$B_{[1,1]}^{Rot} = \frac{1}{2}\left(\mathbb{H}_{Rot}\mathcal{Q}_{[1,1]} - \mathcal{Q}\mathbb{H}_{Rot[1,1]}\right) = 0$$

$$B_{[1,2]}^{Rot} = \frac{1}{2}\left(\mathbb{H}_{Rot}\mathcal{Q}_{[1,2]} - \mathcal{Q}\mathbb{H}_{Rot[1,2]}\right) = \frac{1}{\lambda}\sin\lambda\,(cz - dy)$$

$$B_{[1,3]}^{Rot} = \frac{1}{2}\left(\mathbb{H}_{Rot}\mathcal{Q}_{[1,3]} - \mathcal{Q}\mathbb{H}_{Rot[1,3]}\right) = \frac{1}{\lambda}\sin\lambda\,(dx - bz) \quad (5.19)$$

$$B_{[1,4]}^{Rot} = \frac{1}{2}\left(\mathbb{H}_{Rot}\mathcal{Q}_{[1,4]} - \mathcal{Q}\mathbb{H}_{Rot[1,4]}\right) = \frac{1}{\lambda}\sin\lambda\,(by - cx)$$

This measures the amount of non-commutativity in the rotation. Between them, the E-rotation and the B-rotation are the commutative part of the rotation and the non-commutative part of the rotation respectively.

## Quaternion Rotation

Now, what defines rotation? A rotation is a movement in space which leaves invariant the distance function (norm) of the space. In other words, a vector in the space is the same length after a rotation as it was before the rotation. In even more other words, a point in space is the same distance from the origin after a rotation as it was before the rotation. This invariance of distance from the origin, length of the position vector, is the definition of rotation. Within quaternion space, the distance function (norm) is the determinant of the quaternion. Because the determinant of the quaternion rotation matrix is unity, multiplying a quaternion position vector by the quaternion rotation matrix does not change the determinant of the position vector; it does not change the length of the position vector. Great! Now, does the E-rotation and the B-rotation shown above, (5.16) & (5.17) & (5.18), leave the determinant of the quaternion vector invariant? It does not. These cannot therefore be valid rotations.

Hang on through, within quaternion space, we can rotate in only one 2-dimensional plane at a time. Taking this restricted nature of quaternion rotations by setting two of the three variables to zero in the E-rotation and the B-rotation gives, from (5.18) & (5.19):

$$E^{Rot}_{[1,1]} = t \cos \sqrt{b^2} - x \sin \sqrt{b^2}$$
$$E^{Rot}_{[1,2]} = x \cos \sqrt{b^2} + t \sin \sqrt{b^2}$$
$$E^{Rot}_{[1,3]} = 0$$
$$E^{Rot}_{[1,4]} = 0$$

(5.20)

$$B^{Rot}_{[1,1]} = 0 \qquad B^{Rot}_{[1,2]} = 0$$
$$B^{Rot}_{[1,3]} = 0 \qquad B^{Rot}_{[1,4]} = 0 \qquad (5.21)$$

Done this way, we have the invariance of the distance we require. Of course we do because this is just 2-dimensional rotation and 2-dimensional rotation is commutative. Both bits of the rotation are equal and when we have the total rotation we get a single 2-dimensional rotation.

We conclude the E-rotation and the B-rotation is a valid form of rotation provided we look at only one 2-dimensional plane at a time.

Clifford algebras and double cover:

The quaternions, as well as being a division algebra, are a Clifford algebra, or an even sub-algebra of a Clifford algebra depending upon your point of view. We discuss this is a later chapter, but, for now, it is sufficient for the reader to know that the quaternions are a Clifford algebra and that Clifford algebras are non-commutative algebras.

Within Clifford algebra, double cover rotations are written as:

$$Rot^{-1}.X.Rot \qquad (5.22)$$

Conventionally, the idea of a single rotation matrix rotating a point to two places at once (the double cover so beautifully drawn above) is eschewed by Clifford algebraists. Along with this dislike comes a dislike of producing two different rotations with the same rotation matrix depending upon whether the rotation matrix is placed to the right or to the

## Quaternion Rotation

left of the point to be rotated. Ultimately, this distaste of Clifford algebraists derives from their knowing of only the 2-dimensional rotations we find in the Euclidean complex plane. Clifford algebraists seek to avoid that which they dislike by putting a rotation matrix on both the left and the right of the point to be moved. This produces twice as much rotation as the angle fed into the rotation because there are two rotation matrices, and so the rule is that you must put only half on the angle into the rotation matrices.

Let us consider (5.22) with a quaternion rotation matrix. First of all, the inverse rotation of a quaternion rotation is the same as the rotation – putting $-b$ into the rotation matrix produces the same rotation as putting $+b$ into the rotation matrix. Damn clever these quaternions! Thus we can ignore the inverse and write (5.22) as:

$$\mathbb{H}.X.\mathbb{H} \qquad (5.23)$$

It is only a small wonder that we get $720^0$ worth of rotation when we rotate through $360^0$, but such a rotation will produce only one output for each input.

Clearly, the Clifford algebra rotation is an attempt to define a type of rotation within a non-commutative algebra. Perhaps the non-commutative rotation defined within the quaternions as the E-rotation and the B-rotation is a better definition of non-commutative rotation.

Quaternion multiplication revisited:

There is no logical reason why multiplication within the real numbers is of the form that we take it to be. If we ask 'What is five multiplied by four?', within the real numbers, we

*Quaternions*

conclude that multiplication is just successive addition. Multiplication is just a translation along the real axis. And so, we go into the complex numbers, $\mathbb{C}$, thinking multiplication to be no more than successive addition, but we soon discover that multiplication within the complex plane involves rotation as well as extension. Oh! we will have to rethink multiplication within the real numbers. It does not take much of a rethink; multiplication in the real numbers includes rotation, but rotation in a 1-dimensional space is no more than nothing. We are now at the point of realising that multiplication involves rotation. What is multiplication within a non-commutative division algebra such as the quaternions?

We have so far, along with all other mathematicians, simply assumed, following the example of the 2-dimensional complex plane, $\mathbb{C}$, that multiplication within a non-commutative algebra is simply two matrices multiplied together with 'normal' matrix multiplication. There is no way for us to know what multiplication should be in a non-commutative algebra; all we can do is stumble along hoping to do it correctly. If multiplication is associated with rotation, then perhaps multiplication in a non-commutative algebra like the quaternions should be associated with non-commutative rotation and involve both E-rotation and B-rotation. If this is so, then we would have both the E-product and the B-product. Perhaps we were wrong to unquestioningly assume multiplication in a non-commutative algebra is the same as multiplication in a commutative algebra. Perhaps quaternion multiplication should be of the form:

*Quaternion Rotation*

$$\text{Product}(\mathbb{H}_1, \mathbb{H}_2) = \left\{ \begin{array}{l} \frac{1}{2}(\mathbb{H}_1\mathbb{H}_2 + \mathbb{H}_2\mathbb{H}_1) \\ \frac{1}{2}(\mathbb{H}_1\mathbb{H}_2 - \mathbb{H}_2\mathbb{H}_1) \end{array} \right\} \quad (5.24)$$

We must of course remember the restriction to one 2-dimensional rotation at a time.

Ignoring the half for now, these two expressions are known to particle physicists as the anti-commutator and the commutator; they are usually written as:

$$\{\mathbb{H}_1, \mathbb{H}_2\} = \mathbb{H}_1\mathbb{H}_2 + \mathbb{H}_2\mathbb{H}_1$$
$$[\mathbb{H}_1, \mathbb{H}_2] = \mathbb{H}_1\mathbb{H}_2 - \mathbb{H}_2\mathbb{H}_1 \quad (5.25)$$

Within conventional particle physics, quaternions are not used overtly, but physicists speak of the commutators of operators and the anti-commutators of operators. The operators are 'singled out' 2-dimensional rotations.

So, perhaps multiplication within a non-commutative division algebra is not what it has been unquestioningly assumed to be for the previous 150 years. Perhaps quaternion multiplication is as shown in (5.24). The reader will form their own opinion.

## Chapter 6

# Non-commutative Differentiation

We might want to differentiate quaternion potentials[15]. When we do this, we must take account of the non-commutativity of the quaternions. Differentiation is taking the ratio of an infinitesimal amount of one variable and an infinitesimal amount of another variable, $\frac{\partial y}{\partial x}$. Such a ratio is effectively a product of $\partial y$ and $\frac{1}{\partial x}$. Since quaternions are multiplicatively non-commutative, there is a difference between the quaternion versions of $\partial y \frac{1}{\partial x}$ and $\frac{1}{\partial x} \partial y$.

Before we tackle quaternion differentiation, we need to familiarise the reader with differentiation within division algebras in general. Let us differentiate a Euclidean complex potential. The complex potential is:

$$\mathbb{C}_{Pot} = \begin{bmatrix} \Phi(x,y) & A_x(x,y) \\ -A_x(x,y) & \Phi(x,y) \end{bmatrix} \qquad (6.1)$$

---

[15] Credit for the development of non-commutative differentiation must be shared with David Hestenes for his work on the geometric calculus and with Michael Jack in his paper 'Physical space as a Quaternion Structure'. Jack also cites C.J.Joly 1905 'A Manual of Quaternions'.

We will differentiate it with respect to a complex number:

$$\begin{bmatrix} x & y \\ -y & x \end{bmatrix} \quad (6.2)$$

We separate the differential into its parts. We have:

$$\frac{\partial \begin{bmatrix} \Phi(x,y) & A_x(x,y) \\ -A_x(x,y) & \Phi(x,y) \end{bmatrix}}{\partial \begin{bmatrix} x & y \\ -y & x \end{bmatrix}} = \frac{\partial \begin{bmatrix} \Phi(x,y) & 0 \\ 0 & \Phi(x,y) \end{bmatrix}}{\partial \begin{bmatrix} x & 0 \\ 0 & x \end{bmatrix}}$$

$$+ \frac{\partial \begin{bmatrix} \Phi(x,y) & 0 \\ 0 & \Phi(x,y) \end{bmatrix}}{\partial \begin{bmatrix} 0 & y \\ -y & 0 \end{bmatrix}}$$

$$+ \frac{\partial \begin{bmatrix} 0 & A_x(x,y) \\ -A_x(x,y) & 0 \end{bmatrix}}{\partial \begin{bmatrix} x & 0 \\ 0 & x \end{bmatrix}} + \frac{\partial \begin{bmatrix} 0 & A_x(x,y) \\ -A_x(x,y) & 0 \end{bmatrix}}{\partial \begin{bmatrix} 0 & y \\ -y & 0 \end{bmatrix}}$$

(6.3)

The matrices with elements on only the leading diagonals are isomorphic to real numbers (they are real numbers). We know how to differentiate real numbers with respect to real numbers, and real numbers commute with everything. If we could reduce the separate parts of the differentiation, (6.3), to real numbers, we will be able to do the differentiation. We approach the differentiation a bit at a time. We have:

$$\frac{\partial \begin{bmatrix} \Phi(x,y) & 0 \\ 0 & \Phi(x,y) \end{bmatrix}}{\partial \begin{bmatrix} x & 0 \\ 0 & x \end{bmatrix}} = \begin{bmatrix} \dfrac{\partial \Phi}{\partial x} & 0 \\ 0 & \dfrac{\partial \Phi}{\partial x} \end{bmatrix} \quad (6.4)$$

That bit was easy. We also have:

$$\frac{\partial \begin{bmatrix} \Phi(x,y) & 0 \\ 0 & \Phi(x,y) \end{bmatrix}}{\partial \begin{bmatrix} 0 & y \\ -y & 0 \end{bmatrix}} = \frac{1}{\begin{bmatrix} 0 & 1 \\ -1 & 0 \end{bmatrix}} \frac{\partial \begin{bmatrix} \Phi(x,y) & 0 \\ 0 & \Phi(x,y) \end{bmatrix}}{\partial \begin{bmatrix} y & 0 \\ 0 & y \end{bmatrix}}$$

$$= \begin{bmatrix} 0 & -1 \\ 1 & 0 \end{bmatrix} \begin{bmatrix} \dfrac{\partial \Phi}{\partial y} & 0 \\ 0 & \dfrac{\partial \Phi}{\partial y} \end{bmatrix}$$

$$= \begin{bmatrix} 0 & -\dfrac{\partial \Phi}{\partial y} \\ \dfrac{\partial \Phi}{\partial y} & 0 \end{bmatrix}$$

(6.5)

Because the Euclidean complex numbers are commutative, it does not matter whether we take the imaginary unit variable out of the differential to the right or to the left:

$$\begin{bmatrix} 0 & -1 \\ 1 & 0 \end{bmatrix} \begin{bmatrix} \dfrac{\partial \Phi}{\partial y} & 0 \\ 0 & \dfrac{\partial \Phi}{\partial y} \end{bmatrix} = \begin{bmatrix} \dfrac{\partial \Phi}{\partial y} & 0 \\ 0 & \dfrac{\partial \Phi}{\partial y} \end{bmatrix} \begin{bmatrix} 0 & -1 \\ 1 & 0 \end{bmatrix} \quad (6.6)$$

## Non-Commutative Differentiation

We also have:

$$\frac{\partial \begin{bmatrix} 0 & A_x(x,y) \\ -A_x(x,y) & 0 \end{bmatrix}}{\partial \begin{bmatrix} x & 0 \\ 0 & x \end{bmatrix}} = \begin{bmatrix} 0 & 1 \\ -1 & 0 \end{bmatrix} \frac{\partial \begin{bmatrix} A_x(x,y) & 0 \\ 0 & A_x(x,y) \end{bmatrix}}{\partial \begin{bmatrix} x & 0 \\ 0 & x \end{bmatrix}}$$

$$= \begin{bmatrix} 0 & \frac{\partial A_x}{\partial x} \\ -\frac{\partial A_x}{\partial x} & 0 \end{bmatrix}$$

(6.7)

And finally:

$$\frac{\partial \begin{bmatrix} 0 & A_x(x,y) \\ -A_x(x,y) & 0 \end{bmatrix}}{\partial \begin{bmatrix} 0 & y \\ -y & 0 \end{bmatrix}} = \frac{\begin{bmatrix} 0 & 1 \\ -1 & 0 \end{bmatrix}}{\begin{bmatrix} 0 & 1 \\ -1 & 0 \end{bmatrix}} \begin{bmatrix} \frac{\partial A_x}{\partial y} & 0 \\ 0 & \frac{\partial A_x}{\partial y} \end{bmatrix} \quad (6.8)$$

Pulling it all together gives:

$$\frac{\partial \begin{bmatrix} \Phi(x,y) & A_x(x,y) \\ -A_x(x,y) & \Phi(x,y) \end{bmatrix}}{\partial \begin{bmatrix} x & y \\ -y & x \end{bmatrix}} = \begin{bmatrix} \frac{\partial \Phi}{\partial x} + \frac{\partial A_x}{\partial y} & \frac{\partial A_x}{\partial x} - \frac{\partial \Phi}{\partial y} \\ \frac{\partial \Phi}{\partial y} - \frac{\partial A_x}{\partial x} & \frac{\partial \Phi}{\partial x} + \frac{\partial A_x}{\partial y} \end{bmatrix}$$

$$= \begin{bmatrix} Div & Curl \\ -Curl & Div \end{bmatrix}$$

(6.9)

When we differentiate a quaternion potential, we will get two different differentials depending upon whether we extract the imaginary unit variables to the left, as we have above, or to the right. We will call these differentials $d_L$ for left-differentiation and $d_R$ for right-differentiation. Having calculated both differentials, we form two fields from the differentials which we refer to as the E-field and the B-field. The names reflect the association with the electric and magnetic fields. We have:

$$E = \frac{1}{2}(d_L + d_R)$$
$$B = \frac{1}{2}(d_L - d_R)$$
(6.10)

There is an arbitrariness about whether the B-field is defined as $d_L - d_R$ or $d_R - d_L$. The arbitrariness is what we define to be the positive direction of the B-field.


Differential operators:
Although the correct way to differentiate quaternions is to separate out the four individual parts of the quaternion and differentiate each part with respect to each separate variable (16 separate left differentiations and 16 separate right differentiations), there are many short-cuts which save time and paper.[16] One short-cut is to differentiate the whole quaternion with respect to each separate variable (4 separate

---

[16] The cumbersome but perfectly easy differentiation of a quaternion potential is done in: Dennis Morris: The Physics of Empty Space.

## Non-Commutative Differentiation

left differentiations and 4 separate right differentiations). Another short-cut is the quaternion differentiation operator:

$$\mathbb{H}_{diff} = \begin{bmatrix} \partial_t & -\partial_x & -\partial_y & -\partial_z \\ \partial_x & \partial_t & \partial_z & -\partial_y \\ \partial_y & -\partial_z & \partial_t & \partial_x \\ \partial_z & \partial_y & -\partial_x & \partial_t \end{bmatrix} \quad (6.11)$$

Notice that the minus signs are distributed oppositely to the standard quaternion. This is because we extract the unit imaginary variables from the denominator of the differential and then invert them, as we did above, (6.5), to form the differential. The $\partial_\mu = \dfrac{\partial}{\partial \mu}$ and is just notation. Note: The anti-quaternion differential operator is the standard anti-quaternion with the imaginary signs distributed oppositely.

To differentiate on the left, we simply matrix multiply this differential operator on the left of the quaternion potential with the understanding that $\partial_\mu A_i = \dfrac{\partial A_i}{\partial \mu}$. This works because differentiation is a linear operation. We have the left differential:

$$\mathbb{H}_{diff} Q = \begin{bmatrix} \partial_t & -\partial_x & -\partial_y & -\partial_z \\ \partial_x & \partial_t & \partial_z & -\partial_y \\ \partial_y & -\partial_z & \partial_t & \partial_x \\ \partial_z & \partial_y & -\partial_x & \partial_t \end{bmatrix} \begin{bmatrix} \Phi & A_x & A_y & A_z \\ -A_x & \Phi & -A_z & A_y \\ -A_y & A_z & \Phi & -A_x \\ -A_z & -A_y & A_x & \Phi \end{bmatrix}$$

(6.12)

## Quaternions

This gives (we present only the top row elements of the matrix – that matrix is, of course because of multiplicative closure, a quaternion matrix):

$$\mathbb{H}_{diff} Q_{[1,1]} = \partial_t \Phi + \partial_x A_x + \partial_y A_y + \partial_z A_z$$
$$\mathbb{H}_{diff} Q_{[1,2]} = \partial_t A_x - \partial_x \Phi - \partial_y A_z + \partial_z A_y$$
$$\mathbb{H}_{diff} Q_{[1,3]} = \partial_t A_y + \partial_x A_z - \partial_y \Phi - \partial_z A_x \quad (6.13)$$
$$\mathbb{H}_{diff} Q_{[1,4]} = \partial_t A_z - \partial_x A_y + \partial_y A_x - \partial_z \Phi$$

To differentiate on the right, we simply multiply this differential operator on the right of the quaternion potential with the understanding that $A_i \partial_\mu = \dfrac{\partial A_i}{\partial \mu}$. We have the right differential:

$$Q\,\mathbb{H}_{diff} = \begin{bmatrix} \Phi & A_x & A_y & A_z \\ -A_x & \Phi & -A_z & A_y \\ -A_y & A_z & \Phi & -A_x \\ -A_z & -A_y & A_x & \Phi \end{bmatrix} \begin{bmatrix} \partial_t & -\partial_x & -\partial_y & -\partial_z \\ \partial_x & \partial_t & \partial_z & -\partial_y \\ \partial_y & -\partial_z & \partial_t & \partial_x \\ \partial_z & \partial_y & -\partial_x & \partial_t \end{bmatrix}$$

(6.14)

This gives (we present only the top row elements of the matrix):

$$Q\,\mathbb{H}_{diff[1,1]} = \partial_t \Phi + \partial_x A_x + \partial_y A_y + \partial_z A_z$$
$$Q\,\mathbb{H}_{diff[1,2]} = -\partial_x \Phi + \partial_t A_x - \partial_z A_y + \partial_y A_z$$
$$Q\,\mathbb{H}_{diff[1,3]} = -\partial_y \Phi + \partial_z A_x + \partial_t A_y - \partial_x A_z \quad (6.15)$$
$$Q\,\mathbb{H}_{diff[1,4]} = -\partial_z \Phi - \partial_y A_x + \partial_x A_y + \partial_t A_z$$

## Non-Commutative Differentiation

We form the E-field as half the sum of these left and right differentials:

$$E_{[1,1]} = \frac{1}{2}\left(\mathbb{H}_{\text{diff}}Q_{[1,1]} + Q\,\mathbb{H}_{\text{diff}[1,1]}\right) = \partial_t\Phi + \partial_x A_x + \partial_y A_y + \partial_z A_z$$

$$E_{[1,2]} = \frac{1}{2}\left(\mathbb{H}_{\text{diff}}Q_{[1,2]} + Q\,\mathbb{H}_{\text{diff}[1,2]}\right) = \partial_t A_x - \partial_x\Phi$$

$$E_{[1,3]} = \frac{1}{2}\left(\mathbb{H}_{\text{diff}}Q_{[1,3]} + Q\,\mathbb{H}_{\text{diff}[1,3]}\right) = \partial_t A_y - \partial_y\Phi$$

$$E_{[1,4]} = \frac{1}{2}\left(\mathbb{H}_{\text{diff}}Q_{[1,4]} + Q\,\mathbb{H}_{\text{diff}[1,4]}\right) = \partial_t A_z - \partial_z\Phi$$

(6.16)

We form the B-field as half the the difference of these left and right differentials:

$$B = \frac{1}{2}(d_L - d_R) \qquad (6.17)$$

This is:

$$B_{[1,1]} = \frac{1}{2}\left(\mathbb{H}_{\text{diff}}Q_{[1,1]} - Q\,\mathbb{H}_{\text{diff}[1,1]}\right) = 0$$

$$B_{[1,2]} = \frac{1}{2}\left(\mathbb{H}_{\text{diff}}Q_{[1,2]} - Q\,\mathbb{H}_{\text{diff}[1,2]}\right) = \partial_z A_y - \partial_y A_z$$

(6.18)

$$B_{[1,3]} = \frac{1}{2}\left(\mathbb{H}_{\text{diff}}Q_{[1,3]} - Q\,\mathbb{H}_{\text{diff}[1,3]}\right) = \partial_x A_z - \partial_z A_x$$

$$B_{[1,4]} = \frac{1}{2}\left(\mathbb{H}_{\text{diff}}Q_{[1,4]} - Q\,\mathbb{H}_{\text{diff}[1,4]}\right) = \partial_y A_x - \partial_x A_y$$

Looking at the E-field, the reader will realise that, if we had chosen to reverse the directions of the $A_\mu$ components of the potential, we would have the standard (arbitrary) definition of the electric field:

$$\vec{E} = -Grad\,\Phi - \frac{\partial \vec{A}}{\partial t} \quad (6.19)$$

together with the divergence of that field in 4-dimensional space-time.

Looking at the B-field, the reader will realise that, if we had chosen to reverse the directions of the $A_\mu$ components of the potential, we would have the standard (arbitrary) definition of the magnetic field:

$$\vec{B} = Curl(\vec{A}) \quad (6.20)$$

The reader might now think that the electric field and the magnetic field are quaternion fields. The mathematical squiggles are certainly the same. However, the electromagnetic fields are defined over a 4-dimensional space-time with distance function:

$$d^2 = t^2 - x^2 - y^2 - z^2 \quad (6.21)$$

The quaternion E-field and B-field are defined over quaternion space which has the distance function:

$$d^2 = t^2 + x^2 + y^2 + z^2 \quad (6.22)$$

We have an electromagnetic field over a different kind of space! One wonders if this is the weak nuclear force.

What we have done with the quaternions can equally well be done with the anti-quaternions. Doing the same with the anti-quaternions and superimposing all four fields (that is adding the two B-fields and the two E-fields all together) leads to the electromagnetic tensor over quaternion space.

### The Anti-quaternions:

Differentiating the anti-quaternions using the anti-quaternion differential operator leads to the same quaternion E-field:

$$\begin{aligned} _{Anti}E_{[1,1]} &= \partial_t \Phi + \partial_x A_x + \partial_y A_y + \partial_z A_z \\ _{Anti}E_{[1,2]} &= \partial_t A_x - \partial_x \Phi \\ _{Anti}E_{[1,3]} &= \partial_t A_y - \partial_y \Phi \\ _{Anti}E_{[1,4]} &= \partial_t A_z - \partial_z \Phi \end{aligned} \quad (6.23)$$

And the negative of the quaternion B-field:

$$\begin{aligned} _{Anti}B_{[1,1]} &= 0 \\ _{Anti}B_{[1,2]} &= -\partial_z A_y + \partial_y A_z \\ _{Anti}B_{[1,3]} &= -\partial_x A_z + \partial_z A_x \\ _{Anti}B_{[1,4]} &= -\partial_y A_x + \partial_x A_y \end{aligned} \quad (6.24)$$

### Comments:

It has taken much effort by mathematicians, particularly Clifford algebraists, to figure out how to differentiate within a non-commutative division algebra. Remember that the six $A_3$ division algebras derive alongside the quaternions from the $C_2 \times C_2$ finite group. If we apply the same kind of differentiation to the $A_3$ potentials and superimpose the results, a tensor emerges that splits into the electromagnetic tensor and a symmetric tensor which we take to be the energy-momentum tensor. Further non-commutative differentiation of the $A_3$ potentials gives the Maxwell equations of electromagnetism, but the symmetric field equations which would be gravito-electromagnetism all cancel leaving only the intrinsic curvature of our 4-dimensional space-time to be any kind of force. This curvature corresponds to gravity and to a locally varying $A_3$ phase[17] which induces an affine connection on the emergent locally flat manifold.

The success of non-commutative differentiation leads us to believe that this form of differentiation is the appropriate differentiation to use within a division algebra. With this in mind, we think the non-commutative rotation shown above is the appropriate rotation to use within a non-commutative division algebra.

---

[17] See: Dennis Morris: Upon General Relativity.

Chapter 7

# Double Differentiation

Both the E-field and the B-field are quaternions. Therefore we can differentiate these fields. For historic reasons, the differentials of the E-field and the B-field are written as square brackets and curly brackets. We have:

$$\{E, d\} = \frac{1}{2}\left(d_L E + d_R E\right)$$
$$\{B, d\} = \frac{1}{2}\left(d_L B + d_R B\right)$$
$$[E, d] = \frac{1}{2}\left(d_L E - d_R E\right)$$
$$[B, d] = \frac{1}{2}\left(d_L B - d_R B\right)$$
(7.1)

These are referred to as curly-E, curly-B, straight-E, and straight-B respectively.

We have mentioned above that the direction of the B-field is arbitrary; it is important to use the same arbitrary convention consistently as we have in (6.10) and (7.1).

Because matrices are associative, or by cumbersome calculation, we have:

$$d_L d_R = d_R d_L \qquad (7.2)$$

We have the differential identity:

$$\{B,d\} = \frac{1}{2}\left(d_L d_L - d_L d_R + d_R d_L - d_R d_R\right)$$

$$[E,d] = \frac{1}{2}\left(d_L d_L + d_L d_R - d_R d_L - d_R d_R\right) \qquad (7.3)$$

$$\{B,d\} = [E,d]$$

This differential identity is the homogeneous Maxwell equations over the emergent expectation quaternion space[18].

What we have done with the quaternions can equally well be done with the anti-quaternions.

Superimposition of the quaternions and the anti-quaternions leads not only to the electromagnetic tensor over quaternion space but also to Maxwell's inhomogeneous equations over quaternion space, but that is taking us too far from the core subject of this book.[19]

As we said above, the electromagnetic fields over our 4-dimensional space-time emerge from the superimposition of the six $A_3$ algebras, (3.20) to (3.25).[20]

---

[18] Over our 4-dimensional space-time, this differential identity is the homogeneous Maxwell equations.
[19] See Dennis Morris: The Physics of Empty Space
[20] See Dennis Morris: Upon General Relativity

## Chapter 8

# Connections to Clifford Algebras

The quaternions, as well as being a division algebra, are algebraically isomorphic to a Clifford algebra. Algebraically, they are the Clifford algebra $Cl_{0,2}$ in which we have:

$$\vec{e_1}\vec{e_1} = -1, \quad \vec{e_2}\vec{e_2} = -1, \quad \vec{e_1}\vec{e_2} = -\vec{e_2}\vec{e_1}$$
$$\vec{e_{12}}\vec{e_{12}} = \vec{e_1}\vec{e_2}\vec{e_1}\vec{e_2} = -\vec{e_1}\vec{e_1}\vec{e_2}\vec{e_2} = -1 \quad (8.1)$$

We see $\vec{e_1}$ & $\vec{e_2}$ & $\vec{e_{12}}$ as three square roots of minus unity.

The Clifford algebra $Cl_{0,2}$ is comprised of a scalar (real number) two basis vectors $\vec{e_1}$ & $\vec{e_2}$, and a basis bi-vector $\vec{e_{12}} = \vec{e_1}\vec{e_2}$. It is of the form:

$$a + b\vec{e_1} + c\vec{e_2} + d\vec{e_{12}} \quad (8.2)$$

The distinction between basis vectors and the basis bi-vector is interpretive only. Algebraically, there is no difference between a basis vector and a basis bi-vector. Looking at (8.1), we see that the three imaginary unit variables of the quaternion correspond to the basis vectors and basis bi-vector.

## Quaternions

All the Clifford algebras derive from finite groups of the $C_2 \times C_2 \times ...$ form.[21] There is an 8-dimensional Clifford algebra, $Cl_{0,3}$, which is comprised of two quaternions taken as:

$$Cl_{0,3} \equiv \begin{bmatrix} \mathbb{H}_1 & \mathbb{H}_2 \\ \mathbb{H}_2 & \mathbb{H}_1 \end{bmatrix} \qquad (8.3)$$

The polar form of this matrix is an 8-dimensional division algebra with six imaginary variables which square to minus unity, one imaginary variable which squares to plus unity, and a real scalar. There are 16-dimensional, and in general $2^n$ algebras, which can similarly be seen as formed from quaternions.

The other two 4-dimensional Clifford algebras, $Cl_{2,0}$ & $Cl_{1,1}$ are isomorphic to the $A_3$ division algebras.

---

[21] See: Dennis Morris: The Naked Spinor

## Chapter 9

# Commutation Relations and Lie Groups

The commutation relations of the quaternions are isomorphic as a commutation algebra to the Lie group $SU(2)$. The Lie group $SU(2)$ is a double cover of the Lie group $SO(3)$. Phew! Time for an explanation.

A Lie group is a set of commutation relations. This means it is a set of 'how pairs of non-commutative mathematical objects (usually written as matrices) multiply together'. The matrices must be square and, for a given Lie group, each set of a particular size must have the same multiplicative relations.

The $2\times 2$ matrix representation of $SU(2)$ is:

$$\sigma_1 = \begin{bmatrix} 0 & 1 \\ 1 & 0 \end{bmatrix}, \quad \sigma_2 = \begin{bmatrix} 0 & -i \\ i & 0 \end{bmatrix}, \quad \sigma_3 = \begin{bmatrix} 1 & 0 \\ 0 & -1 \end{bmatrix} \quad (9.1)$$

These are known as the Pauli matrices. The commutator of the matrices is:

$$[\sigma_i, \sigma_j] = \sigma_i \sigma_j - \sigma_j \sigma_i \quad (9.2)$$

For example:

$$[\sigma_1, \sigma_2] = \begin{bmatrix} 0 & 1 \\ 1 & 0 \end{bmatrix}\begin{bmatrix} 0 & -i \\ i & 0 \end{bmatrix} - \begin{bmatrix} 0 & -i \\ i & 0 \end{bmatrix}\begin{bmatrix} 0 & 1 \\ 1 & 0 \end{bmatrix} = 2i\begin{bmatrix} 1 & 0 \\ 0 & -1 \end{bmatrix}$$
$$(9.3)$$

## Quaternions

$$[\sigma_1, \sigma_2] = 2i\sigma_3 \quad (9.4)$$

The essence is that the commutator of any of the matrices is another of the matrices multiplied by a (could be complex) number.

The $3 \times 3$ matrix representation of $SU(2)$ is:

$$J_1 = \frac{1}{\sqrt{2}}\begin{bmatrix} 0 & 1 & 0 \\ 1 & 0 & 1 \\ 0 & 1 & 0 \end{bmatrix}, \quad J_2 = \frac{1}{\sqrt{2}}\begin{bmatrix} 0 & -i & 0 \\ i & 0 & -i \\ 0 & i & 0 \end{bmatrix}$$

$$J_3 = \frac{1}{\sqrt{2}}\begin{bmatrix} 1 & 0 & 0 \\ 0 & 0 & 0 \\ 0 & 0 & -1 \end{bmatrix} \quad (9.5)$$

The reader will often see in the literature that the elements of $SU(2)$ are of the form:

$$\begin{bmatrix} \alpha & -\beta^* \\ \beta & \alpha^* \end{bmatrix} \quad : \{\alpha, \beta\} \in \mathbb{C} \quad : \alpha\alpha^* + \beta\beta^* = 1 \quad (9.6)$$

This means that the commutation relations of $SU(2)$ are succinctly written as this single matrix. If we block substitute the matrix form of a Euclidean complex number, $\mathbb{C}$, into the $SU(2)$ matrix, (9.6), we get:

$$\begin{bmatrix} \alpha & -\beta^* \\ \beta & \alpha^* \end{bmatrix} = \begin{bmatrix} a & b & -c & d \\ -b & a & -d & -c \\ c & d & a & -b \\ -d & c & b & a \end{bmatrix} \quad (9.7)$$

$$: a^2 + b^2 + c^2 + d^2 = 1$$

We have a quaternion with unit determinant (norm). In the polar form of the quaternions, the set of quaternions with unit norm (the set of quaternions unit distance from the origin) is the quaternion rotation matrix. We might take the view that the matrix form of $SU(2)$, (9.6), is 'really' the quaternion rotation matrix in obscure notation.

The Lie group $SU(2)$ is seen as a rotation group very closely related to rotations in the $\mathbb{R}^3$ 3-dimensional space. However, the relationship is accidental. Within the $\mathbb{R}^3$ 3-dimensional space, we write 2-dimensional Euclidean rotations as:

$$\exp\left(\begin{bmatrix} 0 & \theta & 0 \\ -\theta & 0 & 0 \\ 0 & 0 & 0 \end{bmatrix}\right) = \begin{bmatrix} \cos\theta & \sin\theta & 0 \\ -\sin\theta & \cos\theta & 0 \\ 0 & 0 & 1 \end{bmatrix} \quad (9.8)$$

And:

$$\exp\left(\begin{bmatrix} 0 & 0 & \theta \\ 0 & 0 & 0 \\ -\theta & 0 & 0 \end{bmatrix}\right) = \begin{bmatrix} \cos\theta & 0 & \sin\theta \\ 0 & 1 & 0 \\ -\sin\theta & 0 & \cos\theta \end{bmatrix} \quad (9.9)$$

And:

$$\exp\left(\begin{bmatrix} 0 & 0 & 0 \\ 0 & 0 & \theta \\ 0 & -\theta & 0 \end{bmatrix}\right) = \begin{bmatrix} 1 & 0 & 0 \\ 0 & \cos\theta & \sin\theta \\ 0 & -\sin\theta & \cos\theta \end{bmatrix} \quad (9.10)$$

The matrices in the exponential function are called the generators of the rotations; they are taken to be the Lie group $SO(3)$, but the group $SO(3)$ is really the three 2-

dimensional Euclidean rotations in our 4-dimensional space-time.

There are three 2-dimensional orthogonal planes in quaternion space each formed from the real variable and one of the three imaginary variables, $\{a, \hat{ib}\}$, $\{a, \hat{jc}\}$, $\{a, \hat{kd}\}$. These three quaternion planes are not 'fabricated together' as every possible pairing of three variables in the way we see the three 2-dimensional rotations in the spatial part of our 4-dimensional space-time fabricated together as the pairs $\{x, y\}, \{x, z\}, \{y, z\}$.

The way the three quaternion planes fit together is completely different from the 2-dimensional Euclidean rotations we see in our space-time. None-the-less, there are three quaternion pseudo-2-dimensional planes ($C_2 \times C_2$ has three $C_2$ sub-groups) in quaternion space just as there are three 2-dimensional Euclidean rotation planes in our space-time. This coincidence is, in your author's opinion wrongly, taken to be some form of identification between $SU(2)$ and $SO(3)$. Because quaternion rotation is both clockwise and anti-clockwise at the same time, we have twice as much rotation in the quaternions as we have in our space-time Euclidean rotations. The standard mantra is, "$SU(2)$ is a double cover of $SO(3)$" meaning there are three 2-dimensional rotation planes in both groups and that there is twice as much rotation in the quaternions as there is in our space-time. The double cover is associated with the electron having twice as much magnetic moment as would be expected by classical physics.

Of course, the commutation relations of the imaginary variables of a quaternion are (we use the non-matrix notation for ease):

$$\left[\hat{i}, j\right] = \hat{i}j - j\hat{i} = k - (-k) = 2k$$
$$\left[j, k\right] = jk - kj = \hat{i} - (-\hat{i}) = 2\hat{i} \qquad (9.11)$$
$$\left[k, \hat{i}\right] = k\hat{i} - \hat{i}k = j - (-j) = 2j$$

These match the commutation relations of the Pauli matrices other than for the number part.

We see that the quaternions, a full-blown division algebra, have $SU(2)$ within them. Indeed, we might take the view that the quaternion rotation matrix is $SU(2)$, and that invariance under a $SU(2)$ transformation is just rotation in quaternion space. $SU(2)$ is central to quantum physics. The weak nuclear force arises as a consequence of local $SU(2)$ phase variation; we might see this local $SU(2)$ phase variation as just local rotation in quaternion space.

The reader might take the view that division algebras are 'the proper mathematics' and that since we have the $SU(2)$ relations within the quaternion division algebra, we should dispense with the Lie algebra approach as being 'not proper mathematics'. The Lie group $SU(3)$ is a set of commutation relations which is in no division algebra, but there are 8-dimensional division algebras (algebraically isomorphic to Clifford algebras) which have sets of commutation relations. $SU(3)$ is used in Quantum Field Theory to describe the strong force; it is not known if a different set of commutation relations from within an 8-dimensional division algebra would work as well or better in this regard than the $SU(3)$ relations. Such a usage would give six gluons rather than eight gluons; both of which are consistent with observation.

Chapter 10

# The Quaternion Trig. Functions

We begin by differentiating the 2-dimensional Euclidean $\mathbb{C}$ rotation matrix. When differentiating the rotation matrix, we must recall that the argument of the trigonometric functions (the angle) is the imaginary variable. We have:

$$\frac{\partial \begin{bmatrix} \cos\theta & \sin\theta \\ -\sin\theta & \cos\theta \end{bmatrix}}{\partial \begin{bmatrix} 0 & \theta \\ -\theta & 0 \end{bmatrix}} = \begin{bmatrix} 0 & -1 \\ 1 & 0 \end{bmatrix} \begin{bmatrix} -\sin\theta & \cos\theta \\ -\cos\theta & -\sin\theta \end{bmatrix}$$

$$= \begin{bmatrix} \cos\theta & \sin\theta \\ -\sin\theta & \cos\theta \end{bmatrix}$$
(10.1)

We see that the differential of the Euclidean rotation matrix is the Euclidean rotation matrix. This matrix is no more than the exponential function in two dimensions – that's what rotation is after all[22]. In the more traditional notation, we have:

---

[22] Your author must confess that he feels awestruck by the, to him barely perceptible, profundity of this sentence.

$$e^{\hat{i}\theta} = \cos\theta + \hat{i}\sin\theta$$

$$\frac{\partial}{\partial(\hat{i}\theta)}e^{\hat{i}\theta} = e^{\hat{i}\theta} \qquad (10.2)$$

We can rewrite this as:

$$\frac{\partial\left(e^{\begin{bmatrix} 0 & \theta \\ -\theta & 0 \end{bmatrix}}\right)}{\partial\left(\begin{bmatrix} 0 & \theta \\ -\theta & 0 \end{bmatrix}\right)} = e^{\begin{bmatrix} 0 & \theta \\ -\theta & 0 \end{bmatrix}} \qquad (10.3)$$

For two commutative matrices, $\{AB = BA\}$, we have:

$$e^A e^B = e^{AB} \qquad (10.4)$$

For two non-commutative matrices, $\{AB \neq BA\}$, we have:

$$e^A e^B \neq e^{AB} \qquad (10.5)$$

Since the individual variables of a quaternion are not commutative, we might expect a relation analogous to (10.3) not to hold. Let us do with the quaternion rotation matrix what we did above with the 2-dimensional Euclidean rotation matrix.

We will non-commutatively differentiate the quaternion rotation matrix, (5.6), using the quaternion differentiation operator with zero in place of the leading diagonal; the variable on the leading diagonal will produce only zeros anyway. This leads to (recall $\lambda = \sqrt{b^2 + c^2 + d^2}$ ):

## The Quaternion Trig. Functions

$$E_{[1,1]} = \frac{\partial}{\partial b}\frac{b}{\lambda}\sin\lambda + \frac{\partial}{\partial c}\frac{c}{\lambda}\sin\lambda + \frac{\partial}{\partial d}\frac{d}{\lambda}\sin\lambda$$

$$E_{[1,2]} = \frac{\partial}{\partial b}\cos\lambda$$

$$E_{[1,2]} = \frac{\partial}{\partial c}\cos\lambda \qquad (10.6)$$

$$E_{[1,4]} = \frac{\partial}{\partial d}\cos\lambda$$

And to:

$$B_{[1,1]} = 0$$

$$B_{[1,2]} = \frac{\partial}{\partial d}\left(\frac{c}{\lambda}\sin\lambda\right) - \frac{\partial}{\partial c}\left(\frac{d}{\lambda}\sin\lambda\right)$$

$$B_{[1,3]} = \frac{\partial}{\partial b}\left(\frac{d}{\lambda}\sin\lambda\right) - \frac{\partial}{\partial d}\left(\frac{b}{\lambda}\sin\lambda\right) \qquad (10.7)$$

$$B_{[1,4]} = \frac{\partial}{\partial c}\left(\frac{b}{\lambda}\sin\lambda\right) - \frac{\partial}{\partial b}\left(\frac{c}{\lambda}\sin\lambda\right)$$

These are:

$$E_{[1,1]} = \cos\lambda + 2\frac{1}{\lambda}\sin\lambda \qquad (10.8)$$

$$E_{[1,2]} = \frac{b}{\lambda}\sin\lambda$$

$$E_{[1,2]} = \frac{c}{\lambda}\sin\lambda \qquad (10.9)$$

$$E_{[1,4]} = \frac{d}{\lambda}\sin\lambda$$

And:

$$B_{[1,1]} = B_{[1,2]} = B_{[1,3]} = B_{[1,4]} = 0 \qquad (10.10)$$

We see the B-field has disappeared, but that the E-field is the quaternion rotation matrix with an extra $2\dfrac{1}{\lambda}\sin\lambda$ added on to the leading diagonal term.

The anti-quaternions give the same relations.

This extra term on the leading diagonal must be directly related to the non-commutativity of the quaternion rotation matrix.


Second differentials:
The second non-commutative differential, the curly-E, of the quaternion rotation matrix gives:

$$\{E,d\}_{[1,1]} = \cos\lambda + 2\frac{1}{\lambda}\sin\lambda \qquad (10.11)$$

$$\{E,d\}_{[1,2]} = \frac{b}{\lambda}\sin\lambda - 2\frac{b}{\lambda^2}\cos\lambda + 2\frac{b}{\lambda^3}\sin\lambda \quad (10.12)$$

$$\{E,d\}_{[1,3]} = \frac{c}{\lambda}\sin\lambda - 2\frac{c}{\lambda^2}\cos\lambda + 2\frac{c}{\lambda^3}\sin\lambda$$
$$\{E,d\}_{[1,4]} = \frac{d}{\lambda}\sin\lambda - 2\frac{d}{\lambda^2}\cos\lambda + 2\frac{d}{\lambda^3}\sin\lambda \qquad (10.13)$$

*The Quaternion Trig. Functions*

We see that the leading diagonal element, the real part of the quaternion, stays unchanged through this second differentiation but the imaginary parts now change.

The second non-commutative differential straight-E of the quaternion rotation matrix is zero. Since the first non-commutative differential B-field was zero, obviously, the second non-commutative differentials are zero.

Third differentials:

Non-commutatively differentiating the Curly-E field again gives (we place an extra E to the left of the curly brackets to indicate a third differentiation[23]):

$$E\{E,d\}_{[1,1]} = \cos \lambda + 4\frac{1}{\lambda}\sin \lambda \qquad (10.14)$$

$$E\{E,d\}_{[1,2]} = \frac{b}{\lambda}\sin \lambda - 2\frac{b}{\lambda^2}\cos \lambda + 2\frac{b}{\lambda^3}\sin \lambda$$

$$E\{E,d\}_{[1,3]} = \frac{c}{\lambda}\sin \lambda - 2\frac{c}{\lambda^2}\cos \lambda + 2\frac{c}{\lambda^3}\sin \lambda$$

$$(10.15)$$

$$E\{E,d\}_{[1,4]} = \frac{d}{\lambda}\sin \lambda - 2\frac{d}{\lambda^2}\cos \lambda + 2\frac{d}{\lambda^3}\sin \lambda$$

$$(10.16)$$

---

[23] Your author apologises for this poor notation. We can't use an upper case D without confusing this with covariant differentiation. The historic Curly bracket notation perhaps needs changing.

The real part has changed this time, but the imaginary parts remain unchanged. The B-field of Curly-E is zero.

Fourth differentials:
The E-field of the E-field of Curly-E is:

$$EE\{E,d\}_{[1,1]} = \cos \lambda + 4\frac{1}{\lambda}\sin \lambda$$

$$EE\{E,d\}_{[1,2]} = \frac{b}{\lambda}\sin \lambda - 4\frac{b}{\lambda^2}\cos \lambda + 4\frac{b}{\lambda^3}\sin \lambda$$

$$EE\{E,d\}_{[1,3]} = \frac{c}{\lambda}\sin \lambda - 4\frac{c}{\lambda^2}\cos \lambda + 4\frac{c}{\lambda^3}\sin \lambda$$

$$EE\{E,d\}_{[1,4]} = \frac{d}{\lambda}\sin \lambda - 4\frac{d}{\lambda^2}\cos \lambda + 4\frac{d}{\lambda^3}\sin \lambda$$

(10.17)

The B-field of the E-field of Curly-E is zero.

The reader can examine the results and see the pattern for themselves. Your author wonders if this is connected to the Baker-Campbell-Hausdorff formula. There's a PhD thesis for someone.

Orthogonality:
Mathematicians tell us that two functions, $f(x) \& g(x)$, are orthogonal if:

$$\langle f(x), g(x) \rangle = \int_{-a}^{+a} dx \; f^*(x) g(x) = 0 \quad (10.18)$$

The justification of this is that the given integration:

a) gives a real number that is positive definite or zero
b) is symmetrical
c) is bilinear

These are properties that are used to define an inner product of vectors, or, more technically, to define the inner product over a linear space (also called a vector space). We are familiar with the inner product of a division algebra as a measure of the angle between two vectors within that division algebra, and all division algebras are automatically linear spaces, but not all linear spaces are division algebras. Functions form an infinite dimensional linear space which is not a division algebra. Can an infinite dimensional linear space which is not a division algebra really have any meaningful concept of angle and thus of inner product?

Many people are unhappy about the concept of infinite dimensional linear spaces, but this concept is an extremely useful concept when it comes to solving differential equations. If the reader is unhappy at the concept of infinite dimensional linear spaces, then perhaps the reader will be more settled if she sees these infinite dimensional linear spaces as merely a tool. We do not need the concepts to be true; we need only that the concepts give the correct answers to our differential equations; they do.

An example of two orthogonal functions is the two trigonometric functions of the Euclidean complex plane:

$$\int_{-a}^{+a} dx \ \sin\left(\frac{k\pi x}{a}\right)\cos\left(\frac{m\pi x}{a}\right) = 0 \qquad (10.19)$$

$$k \neq m$$

This is true for all values of the limit $a$.

*Quaternions*

There are three variables within the four quaternion trigonometric functions. Those trigonometric functions are:

$$NuA = \cos\left(\sqrt{b^2 + c^2 + d^2}\right)$$
$$NuB = \frac{b}{\sqrt{b^2 + c^2 + d^2}} \sin\left(\sqrt{b^2 + c^2 + d^2}\right)$$
$$NuC = \frac{c}{\sqrt{b^2 + c^2 + d^2}} \sin\left(\sqrt{b^2 + c^2 + d^2}\right) \quad (10.20)$$
$$NuD = \frac{d}{\sqrt{b^2 + c^2 + d^2}} \sin\left(\sqrt{b^2 + c^2 + d^2}\right)$$

We have:

$$\int_{-a}^{+a} db \; NuA.NuB = 0$$
$$\int_{-a}^{+a} dc \; NuA.NuC = 0 \quad (10.21)$$
$$\int_{-a}^{+a} dd \; NuA.NuD = 0$$

Your author apologises for the notational clash in the last of the above. The variable with respect to which we integrate has got to match the imaginary trigonometric function:

$$\int_{-a}^{+a} db \; NuA.NuC \neq 0 \quad (10.22)$$

We also have:

$$\int_{-a}^{+a} db \ NuB.NuC = 0$$

$$\int_{-a}^{+a} db \ NuB.NuD = 0 \qquad (10.23)$$

$$\int_{-a}^{+a} dc \ NuC.NuD = 0$$

Again the variable with respect to which we integrate has to match one of the trigonometric functions; either will do, by the similar nature of the trigonometric functions. However:

$$\int_{-a}^{+a} db \ NuC.NuD \neq 0 \qquad (10.24)$$

Given the above, it is hardly surprising that:

$$\int_{-a}^{+a} db \ NuB.NuC.NuD = 0 \qquad (10.25)$$

This is true for integration with respect to any of the three variables, by the similar nature of the trigonometric functions. However:

$$\int_{-a}^{+a} db \ NuA.NuC.NuD \neq 0$$

$$\int_{-a}^{+a} dc \ NuA.NuC.NuD = 0 \qquad (10.26)$$

Of course:

$$\int_{-a}^{+a} db \ NuA.NuB.NuC.NuD = 0 \qquad (10.27)$$

It follows that:

$$\int_{b=-a}^{b=+a} db \int_{c=-a}^{c=+a} dc \int_{d=-a}^{d=+a} dd \ NuA.NuB.NuC.NuD = 0$$
(10.28)

And so, by the standard definition, in any of its many interpretations, the quaternion trigonometric functions are orthogonal.

Given that trigonometric functions are projections from the unit sphere of a geometric space on to each of the axes of that space and that the axes are orthogonal, it is hardly surprising that the quaternion trigonometric functions are 'at right-angles' to each other. The same is , of course, in a geometric way, true of any set of trigonometric functions within any division algebra.

# Chapter 11

# The Quaternion Inner Product

The Euclidean inner product:
Within the 2-dimensional Euclidean complex numbers, we have two normalised complex numbers in polar form:

$$\begin{bmatrix} \cos\theta & -\sin\theta \\ \sin\theta & \cos\theta \end{bmatrix} \begin{bmatrix} \cos\phi & \sin\phi \\ -\sin\phi & \cos\phi \end{bmatrix} = \begin{bmatrix} \cos(\theta-\phi) & \sin(\theta-\phi) \\ -\sin(\theta-\phi) & \cos(\theta-\phi) \end{bmatrix} \quad (11.1)$$

In normalised Cartesian form, this is:

$$\frac{1}{\sqrt{a^2+b^2}} \begin{bmatrix} a & -b \\ b & a \end{bmatrix} \frac{1}{\sqrt{c^2+d^2}} \begin{bmatrix} c & d \\ -d & c \end{bmatrix} = \frac{1}{\sqrt{a^2+b^2}\sqrt{c^2+d^2}} \begin{bmatrix} ac+bd & ad-bc \\ -(ad-bc) & ac+bd \end{bmatrix} \quad (11.2)$$

By putting these equal, we see that the inner product of two complex numbers is the angle between those two complex numbers. In more usual notation:

$$\cos(\theta-\phi) = \frac{ac+bd}{\sqrt{a^2+b^2}\sqrt{c^2+d^2}} \quad (11.3)$$

An inner product is a measure of the angle between two vectors and is equal to the leading diagonal trigonometric function of the space of the vectors. It follows that only division algebras can hold an inner product because only division algebras hold angles. This rather makes the concept of Hilbert space a little awry.

Traditionally, the quaternion inner product is seen as being a measure of the angle between two quaternions and is put equal to the cosine function as:

$$\cos\theta = \frac{ae+bf+cg+dh}{\sqrt{a^2+b^2+c^2+d^2}\sqrt{e^2+f^2+g^2+h^2}} \quad (11.4)$$

The left-hand side of this is nonsense. We do not have 2-dimensional angles within quaternion space; we have 4-dimensional quaternion angles in quaternion space. The right-hand side of this is correct. Multiplying two normalised quaternion matrices in Cartesian form together as we did above with the Euclidean complex numbers, (11.2), gives exactly the right-hand side of (11.4) as the element on the leading diagonal, the real part of the quaternion, regardless of the order of multiplication.

The correct left-hand side of the quaternion inner product is given by multiplying the conjugate of one quaternion rotation matrix with another quaternion rotation matrix. The real part of the product is unaffected by the order of the multiplication, and so the inner product is well defined.

That inner product is, with $\lambda = \sqrt{\theta^2 + \phi^2 + \varphi^2}$ and $\kappa = \sqrt{\eta^2 + \psi^2 + \omega^2}$ :

*The Quaternion Inner Product*

$$\cos \lambda \cos \kappa + \frac{(\theta\eta + \phi\psi + \varphi\omega)\sin \lambda \sin \kappa}{\lambda \kappa}$$
$$= \frac{ae + bf + cg + dh}{\sqrt{a^2 + b^2 + c^2 + d^2}\sqrt{e^2 + f^2 + g^2 + h^2}} \quad (11.5)$$

The trigonometric part of the above, (11.5), reduces to the almost familiar 2-dimension Euclidean trigonometric relation:

$$\cos\sqrt{\theta^2}\cos\sqrt{\eta^2} + \frac{\theta\eta}{\sqrt{\theta^2}\sqrt{\eta^2}}\sin\sqrt{\theta^2}\sin\sqrt{\eta^2} \quad (11.6)$$

Which we might put equal to $= \cos\left(\sqrt{\theta^2} - \sqrt{\eta^2}\right)$

Now, because the quaternions are a division algebra, when we multiply two quaternions together, even if one of the quaternions is in its conjugate form, we get a quaternion. Similarly, if we multiply together two quaternion rotation matrices together, we get a quaternion rotation matrix. The order of multiplication is irrelevant for the element on the leading diagonal, the real part, of the quaternion. A quaternion rotation matrix has a $\cos(\chi)$ on the leading diagonal. We therefore have the trigonometric identity:

$$\cos \chi = \cos \lambda \cos \kappa + \frac{(\theta\eta + \phi\psi + \varphi\omega)\sin \lambda \sin \kappa}{\lambda \kappa}$$
$$(11.7)$$

$$\chi \sim \sqrt{l^2 + m^2 + n^2} \quad (11.8)$$

### Trigonometric identities:

The trigonometric identities of the Euclidean complex numbers are derived from either the determinant of a rotation matrix or from multiplying together two rotation matrices such as (11.1). The determinant of a quaternion rotation matrix is unity, of course, and so we have:

$$\cos^2 \lambda + \sin^2 \lambda = 1 \tag{11.9}$$

Trigonometric identities analogous to (11.1) are improperly defined for quaternions using a simple product of two quaternion rotation matrices because the quaternions are non-commutative. Instead, we use the E-rotation and B-rotation shown above, (5.16)~(5.19) to define the trigonometric identities.

We multiply two quaternion rotation matrices together and form the E-rotation and the B-rotation. We get:

$$E_{[1,1]}^{Rot} = \cos \lambda \cos \kappa + \frac{(bf + cg + dh)}{\lambda \kappa} \sin \lambda \sin \kappa$$

$$E_{[1,2]}^{Rot} = \frac{f}{\kappa} \cos \lambda \sin \kappa - \frac{b}{\lambda} \sin \lambda \cos \kappa$$

$$E_{[1,3]}^{Rot} = \frac{g}{\kappa} \cos \lambda \sin \kappa - \frac{c}{\lambda} \sin \lambda \cos \kappa$$

$$E_{[1,4]}^{Rot} = \frac{h}{\kappa} \cos \lambda \sin \kappa - \frac{d}{\lambda} \sin \lambda \cos \kappa$$

$$(11.10)$$

Wherein $\lambda = \sqrt{b^2 + c^2 + d^2}$ and $\kappa = \sqrt{f^2 + g^2 + h^2}$.

We see that the LHS of the trigonometric identity, (11.7), has emerged from the non-commutative rotation. We

*The Quaternion Inner Product*

similarly have the quaternion trigonometric identities, based on non-commutative rotation:

$$\cos \chi = \cos \lambda \cos \kappa - \frac{(bf + cg + dh)}{\lambda \kappa} \sin \lambda \sin \kappa$$

$$\frac{\chi_1 \sin \chi}{\chi} = \frac{f}{\kappa} \cos \lambda \sin \kappa - \frac{b}{\lambda} \sin \lambda \cos \kappa$$

$$\frac{\chi_2 \sin \chi}{\chi} = \frac{g}{\kappa} \cos \lambda \sin \kappa - \frac{c}{\lambda} \sin \lambda \cos \kappa$$

$$\frac{\chi_3 \sin \chi}{\chi} = \frac{h}{\kappa} \cos \lambda \sin \kappa - \frac{d}{\lambda} \sin \lambda \cos \kappa$$

(11.11)

$\chi_i$ are the different variables under the square root sign of $\chi$, (11.8).

Looking at the separate quaternion trigonometric functions, this is what we might have guessed analogously to the 2-dimensional Euclidean trigonometric functions.

As well as the E-rotation, non-commutative rotation gives us a B-rotation. Unfortunately, your author has no idea what-so-ever what to do with the non-commutative B-rotation. It looks like some kind of curl.

The B-rotation is:

$$B^{Rot}_{[1,1]} = 0$$
$$B^{Rot}_{[1,2]} = \frac{(ch - dg)}{\kappa \lambda} \sin \lambda \sin \kappa \qquad (11.12)$$

$$B_{[1,3]}^{Rot} = \frac{(df - bh)}{\kappa \lambda} \sin \lambda \sin \kappa$$
$$B_{[1,4]}^{Rot} = \frac{(bg - cf)}{\kappa \lambda} \sin \lambda \sin \kappa$$
(11.13)

We must remember that quaternion rotation can happen only in a single 2-dimensional plane. With this in mind, the B-rotations would be zero.

Chapter 12

# The Spinors of Physicists

The Schrödinger equation has a single Euclidean complex number as a variable; it describes a non-relativistic 'spinless' fermion (electron). There is no such thing as a 'spinless' fermion (electron), and a complete description of a non-relativistic fermion (electron) is the Pauli-Schrödinger equation which has a pair of ordered complex numbers for a variable. Such a pair of complex numbers is called a spinor by physicists.[24]

The Dirac equation of QFT uses four ordered complex numbers as a variable, but, for particles moving at the speed of light, the Dirac equation splits into two uncoupled equations each using an ordered pair of complex numbers as a variable.

The nature of these pairs of complex numbers, spinors, is:

$$\psi = \begin{bmatrix} a + \hat{i}b \\ c + \hat{i}d \end{bmatrix} \qquad (12.1)$$

The norm of these objects is calculated by the physicists as:

---

[24] There are differing views over whether or not the single complex number of the Schrödinger equation is a spinor. Some say it is a spinor because in is in the Schrödinger equation; others say it is not a spinor because it does not have double cover. The dispute is no more than nomenclature.

$$\begin{bmatrix} a-\hat{i}b & c-\hat{i}d \end{bmatrix} \begin{bmatrix} a+\hat{i}b \\ c+\hat{i}d \end{bmatrix} = a^2 + b^2 + c^2 + d^2 \quad (12.2)$$

It is quite striking that these spinors, which are utterly central to quantum physics, should have their length measured in quaternion space. Perhaps they are secretly quaternions.

Let us see if they might be quaternions. Above, (9.7), we introduced the matrix:

$$\begin{bmatrix} \alpha & -\beta^* \\ \beta & \alpha^* \end{bmatrix} = \begin{bmatrix} a & b & -c & d \\ -b & a & -d & -c \\ c & d & a & -b \\ -d & c & b & a \end{bmatrix} \quad (12.3)$$

This says that the four variables of a quaternion can be written as the four variables of a pair of complex numbers:

$$\alpha = \begin{bmatrix} a & b \\ -b & a \end{bmatrix} \ \& \ \beta = \begin{bmatrix} c & d \\ -d & c \end{bmatrix} \quad (12.4)$$

The sign of the $c$ variable is swapped from the usual quaternion notation, but this is of no consequence. Does this imply that we can denote the quaternions as the spinor?:

$$\begin{bmatrix} \alpha \\ \beta \end{bmatrix} \quad : \{\alpha, \beta\} \in \mathbb{C} \quad (12.5)$$

We cannot say the quaternions are the same thing as the spinor because the spinor is traditionally just an element of a complex linear space and not an element of a division algebra, but, certainly, we can write the quaternions as a pair of complex numbers.

There are many, possibly infinitely many, ways of writing quaternions. We choose the matrix notation in this book because we feel that, although it is cumbersome, it is easily seen through. It is not immediately apparent from the conventional quaternion notation that quaternion multiplication is just matrix multiplication and that the quaternions derive from the finite group $C_2 \times C_2$; this is more easily seen using matrix notation. Most physicists have not realised that their spinor notation has much in common with quaternions; this is an example of how some notations can obscure what is really happening. The reader might prefer to write quaternions as a pair of complex numbers. There is nothing mathematically wrong in doing this, but also there is also nothing wrong in a literary sense with writing poetry in Egyptian hieroglyphics.

We calculate the norm of the quaternion $a + \hat{i}b + \hat{j}c + \hat{k}d$ by multiplying it by its conjugate, $a - \hat{i}b - \hat{j}c - \hat{k}d$, to obtain the norm as $a^2 + b^2 + c^2 + d^2$. To obtain this from the pair of complex number notation, (12.4) & (12.5), we need to form the conjugate as:

$$\begin{bmatrix} \alpha^* & \beta^* \end{bmatrix} \quad (12.6)$$

This gives the norm as:

$$n^2 = \begin{bmatrix} \alpha^* & \beta^* \end{bmatrix} \begin{bmatrix} \alpha \\ \beta \end{bmatrix} \quad (12.7)$$

Which is, as we said above, of course, exactly how physicists handle these 'spinor' pairs of complex numbers.

The given conjugate, (12.6), does not appear to be equal to the quaternion conjugate because of a sign difference on the real part (the $c$ variable) of the second complex number, but this arises from the notation and is of no mathematical importance.

Spinors and division algebras:

The quaternions are a division algebra. Spinors written as ordered pairs of complex numbers are only a complex Hilbert space. If we are prepared to leave spinors as being no more than a complex Hilbert space, then we cannot say that spinors are really quaternions.

A complex Hilbert space is a complex linear space with an inner product. A linear space satisfies seven of the thirteen division algebra axioms, but it has no multiplication operation – no way of multiplying two ordered pairs of complex numbers together. If we are to take these pairs of complex numbers to be quaternions, we need a way of multiplying two of them together which satisfies the axioms of a division algebra (multiplicative distributivity over addition, absence of zero divisors, multiplicative inverses etc.). We are free to choose any suitable multiplication operation since the Hilbert space does not have one; the quaternion multiplication operation will do, and it is linear multiplication. We know this satisfies the axioms of a division algebra.

The product of two quaternions is, we give the top row of the matrix only:

*The Spinors of Physicists*

$$QQ_{[1,1]} = ae - bf - cg - dh$$
$$QQ_{[1,2]} = af + be + ch - dg$$
$$QQ_{[1,3]} = ag - bh + ce + df \quad (12.8)$$
$$QQ_{[1,4]} = ah + bg - cf + de$$

Formally: We define multiplication of these ordered pairs of complex numbers to be:

$$\begin{bmatrix} \alpha \\ \beta \end{bmatrix} \times \begin{bmatrix} \gamma \\ \delta \end{bmatrix} = \begin{bmatrix} a+ib \\ c+id \end{bmatrix} \times \begin{bmatrix} e+if \\ g+ih \end{bmatrix}$$
$$= \begin{bmatrix} ae - bf - cg - dh + i(af + be + ch - dg) \\ ag - bh + ce + df + i(ah + bg - cf + de) \end{bmatrix} \quad (12.9)$$

The four parts of this product correspond to the four parts of a quaternion.

There is a 'kind of multiplication operation with spinors called the outer product. Strictly speaking, a Hilbert space, complex or otherwise, has no multiplication operation and so a Hilbert space has no outer product. The physicists are 'fudging' the maths to get an outer product of two spinors. That outer product is:

$$|\phi\rangle\langle\psi| = \begin{bmatrix} a+\hat{i}b \\ c+\hat{i}d \end{bmatrix} \begin{bmatrix} e+\hat{i}f & g+\hat{i}h \end{bmatrix}$$
$$= \begin{bmatrix} ae - bf + \hat{i}(af + be) & ag - bh + \hat{i}(ah + bg) \\ ce - df + \hat{i}(cf + de) & cg - dh + \hat{i}(ch + dg) \end{bmatrix}$$
$$(12.10)$$

## Quaternions

We see, (12.9), that, other than a couple of signs, this is the quaternion multiplication we defined above.

Addition is already defined on the Hilbert space as:

$$\begin{bmatrix} \alpha \\ \beta \end{bmatrix} + \begin{bmatrix} \gamma \\ \delta \end{bmatrix} = \begin{bmatrix} \alpha + \gamma \\ \beta + \delta \end{bmatrix} \quad (12.11)$$

Multiplication by a complex number is:

$$(a + \hat{i}b)\begin{bmatrix} e + \hat{i}f \\ g + \hat{i}h \end{bmatrix} = \begin{bmatrix} ae - bf + \hat{i}(af + be) \\ ag - bh + \hat{i}(ah + bg) \end{bmatrix} \quad (12.12)$$

In quaternion form, this is:

$$\begin{bmatrix} a & b & 0 & 0 \\ -b & a & 0 & 0 \\ 0 & 0 & a & -b \\ 0 & 0 & b & a \end{bmatrix} \begin{bmatrix} e & f & g & h \\ -f & e & -h & g \\ -g & h & e & -f \\ -h & -g & f & e \end{bmatrix} =$$

$$\begin{bmatrix} ae - bf & af + be & ag - bh & ah + bg \\ -(af + be) & ae - bf & -(ah + bg) & ag - bh \\ -(ag - bh) & ah + gb & ae - bf & -(af + be) \\ -(ah + bg) & -(ag - bh) & af + be & ae - bf \end{bmatrix}$$

(12.13)

The spinor view holds the conjugate to be a dual space to the spinor space. Within a division algebra, there is not a dual space in the usual sense, but every quaternion has a conjugate. We could take the view that quaternions are associated with clockwise rotation and conjugate quaternions are associated with anti-clockwise rotation and

## The Spinors of Physicists

so the division algebras splits into a space and a dual space, but conjugate quaternions are just as much quaternions as any other quaternion, and so the splitting into two spaces is artificial; we do not need the concept of a dual space to form the inner product; so why would we want it.

The spinors inner product is such that:

$$\langle \alpha, \beta \rangle = \langle \beta, \alpha \rangle^* \qquad (12.14)$$

This is:

$$\begin{bmatrix} a - \hat{i}b & c - \hat{i}d \end{bmatrix} \begin{bmatrix} e + \hat{i}f \\ g + \hat{i}h \end{bmatrix} =$$
$$ae + bf + cg + dh + \hat{i}(af - be + ch - dg)$$
$$\begin{bmatrix} e - \hat{i}f & g - \hat{i}h \end{bmatrix} \begin{bmatrix} a + \hat{i}b \\ c + \hat{i}d \end{bmatrix} = \qquad (12.15)$$
$$ae + bf + cg + dh - \hat{i}(af - be + ch - dg)$$

This is not quite what we get if we reverse the order of multiplication of a quaternion and a conjugate quaternion; there are a couple of signs different.

$$Q^*Q_{[1,1]} = ae + bf + cg + dh$$
$$Q^*Q_{[1,2]} = af - be - ch + dg$$
$$QQ^*_{[1,1]} = ae + bf + cg + dh \qquad (12.16)$$
$$QQ^*_{[1,2]} = af - be + ch - dg$$

## Quaternions

Such difference will be 'lost' when the physicist calculates the probability as the square of the inner product and normalises the answer.

With these definitions, we run through the axioms of a division algebra listed above (circa (2.1)) and we find that these ordered pairs of complex numbers satisfy all the division algebra axioms and are thus a division algebra. They have the same norm as the quaternions (or antiquaternions). If two division algebras have the same norm, they are algebraically isomorphic (the same algebra). These pairs of complex numbers are an obscure way of writing a quaternion.

So, we might refer to the spinors with a multiplication operation as 'multiplicative spinors' or 'upper class spinors' or something. Whatever we call them, with the multiplication operation, spinors are a division algebra.


### The inner product:
To a physicist, the inner product of a pair of spinors is just a way of calculating expectation values of variables and the probabilities of finding a particular eigenvalue for a variable.

Within a division algebra, an inner product is the measure of the angle between two vectors expressed as the argument of the leading trigonometric function.

Division algebras have angles and leading trigonometric functions that can take those angles as arguments. Hilbert spaces have neither angles nor trigonometric functions,

*The Spinors of Physicists*

which leads one to question what an inner product really is in a Hilbert space. None-the-less, let us proceed.

The quaternion inner product of two quaternions is given as the real part (leading diagonal) of:

$$IP_{\mathbb{H}} = \begin{bmatrix} a & -b & -c & -d \\ b & a & d & -c \\ c & -d & a & b \\ d & c & -b & a \end{bmatrix} \begin{bmatrix} e & f & g & h \\ -f & e & -h & g \\ -g & h & e & -f \\ -h & -g & f & e \end{bmatrix}$$

(12.17)

This product is:

$$\begin{aligned} IP_{\mathbb{H}[1,1]} &= ae + bf + cg + dh \\ IP_{\mathbb{H}[1,2]} &= af - be - ch + dg \\ IP_{\mathbb{H}[1,3]} &= ag + bh - ce - df \\ IP_{\mathbb{H}[1,4]} &= ah - bg + cf - de \end{aligned}$$

(12.18)

As well as the inner product, the real variable in this product, we have three imaginary parts. Strictly speaking, only the real part is the inner product, but let us be a little lax with the nomenclature and refer to all four parts as the inner product.

The conventional inner product of two spinors is:

$$IP_{spinors} = \begin{bmatrix} a - \hat{i}b & c - \hat{i}d \end{bmatrix} \begin{bmatrix} e + \hat{i}f \\ g + \hat{i}h \end{bmatrix} = \quad (12.19)$$
$$ae + bf + cg + dh + i(af - be + ch - dg)$$

*Quaternions*

As well as a real part, this inner product has an imaginary part; it's not very good maths, but let us proceed on the understanding that this is acceptable. We notice that the real part of this inner product, (12.19), is the same as the real part of the quaternion inner product, (12.18), but that the imaginary part of (12.19) differs from the second part of the quaternion inner product (12.18) by a couple of signs. There are also two more parts to the quaternion inner product, of course.

We can swap the direction of the $c$ variable in the spinors; this would work, but we choose instead to reverse the sign of the $d$ variable and write:

$$\mathbb{H} = \begin{bmatrix} a & b & c & d \\ -b & a & -d & c \\ -c & d & a & -b \\ -d & -c & b & a \end{bmatrix} \simeq \begin{bmatrix} \pm(a+ib) \\ \pm(c-id) \end{bmatrix}_{(\mathbb{H}:\times)}$$

(12.20)

We can choose any sign we like for the two parts of the spinor. The inner product of two spinors of the form shown now matches the real and first imaginary part of the quaternion, (12.18):

$$IP_{spinors} = \begin{bmatrix} a-\hat{i}b & c+\hat{i}d \end{bmatrix} \begin{bmatrix} \pm(e+\hat{i}f) \\ \pm(g-\hat{i}h) \end{bmatrix} = \quad (12.21)$$
$$ae+bf+cg+dh+i(af-be-ch+dg)$$

We can now say that the spinor of the form shown in (12.21) is a quaternion with the proviso that we have imposed the quaternion multiplication operation (matrix multiplication) upon the spinors to convert them from a Hilbert space to a division algebra. We are still stuck with the inner product having an imaginary part and the two absent parts of the quaternion product, but let us proceed with hope and faith.

Within the Euclidean complex numbers, we have the calculation of the inner product as:

$$\frac{1}{\sqrt{a^2+b^2}\sqrt{c^2+d^2}}\begin{bmatrix} a & -b \\ b & a \end{bmatrix}\begin{bmatrix} c & d \\ -d & c \end{bmatrix} =$$
$$\frac{1}{\sqrt{a^2+b^2}\sqrt{c^2+d^2}}\begin{bmatrix} ac+bd & ad-bc \\ -(ad-bc) & ac+bd \end{bmatrix} \quad (12.22)$$
$$= \begin{bmatrix} \cos\theta & \sin\theta \\ -\sin\theta & \cos\theta \end{bmatrix}$$

We see that the real part of (12.22) is just the magnitude of the cosine of the angle between the vectors. We see also that the imaginary part is the just the magnitude of the sine of the angle between the vectors. Both the real part and the imaginary part of the product of a complex number and the conjugate of a complex number measure the size of the angle between the two vectors in the complex plane, but they measure it in different ways through different trigonometric functions. With the quaternion inner product calculation, the real part and the three imaginary parts are also measures of the size of the quaternion angle between two vectors measured through different trigonometric functions.

Looking at (12.18) & (12.19), we see that the Euclidean complex product, (12.19), is just the sum of two different measures of the angle between two vectors and that the quaternion inner product, (12.18), is just four measures of the same angle. We need only one measure; in particular, the three imaginary quaternion trigonometric functions are basically the same; we can discard the third and fourth imaginary parts of the quaternion inner product, (12.18), and lose nothing. We could discard the second imaginary part and lose nothing, but we will keep it for consistency with the spinor mathematics.

What do physicists do with the inner product of a pair of spinors? They square it and call the result a probability. We have:

$$\left|IP_{spinors}\right|^2 = \left|\begin{bmatrix} a-\hat{i}b & c+\hat{i}d \end{bmatrix}\begin{bmatrix} e+\hat{i}f \\ g-\hat{i}h \end{bmatrix}\right|^2 =$$

$$\left|ae+bf+cg+dh+i\left(af-be-ch+dg\right)\right|^2$$

$$= \left(ae+bf+cg+dh\right)^2 + \left(af-be-ch+dg\right)^2$$

$$= Probability$$

(12.23)

The result of this is slightly different from what the result would have been if we had not reversed the sign of the $ch$ and the $dg$ variable by putting a minus sign before the imaginary part of the second spinor component.

We see that the probability, (12.23), is the sum of the squares of two different measures of the angle between two quaternion vectors.

Including the normalisation of the Cartesian forms of the vectors gives:

*The Spinors of Physicists*

$$\cos\sqrt{\phi^2 + \theta^2 + \varphi^2} = \frac{ae + bf + cg + dh}{\sqrt{a^2 + b^2 + c^2 + d^2}\sqrt{e^2 + f^2 + g^2 + h^2}}$$

$$\frac{\phi\sin\sqrt{\phi^2 + \theta^2 + \varphi^2}}{\sqrt{\phi^2 + \theta^2 + \varphi^2}} = \frac{af - be - ch + dg}{\sqrt{a^2 + b^2 + c^2 + d^2}\sqrt{e^2 + f^2 + g^2 + h^2}}$$

(12.24)

Adding two different measures of the same thing produces a measure of that thing, albeit a complicated measure, which can be adjusted by scaling (normalisation). Physicists always normalise probability so that total probability must be unity, and so we see a reconciliation between the real and imaginary spinor inner product and the purely real quaternion inner product. They are really just the same thing normalised differently.

Anti-quaternions and anti-spinors:
The inner product of the anti-quaternions is:

$$AntiQ - IP_{\mathbb{H}[1,1]} = ae + bf + cg + dh$$
$$AntiQ - IP_{\mathbb{H}[1,2]} = af - be + ch - dg$$
$$AntiQ - IP_{\mathbb{H}[1,3]} = ag - bh - ce + df$$
$$AntiQ - IP_{\mathbb{H}[1,4]} = ah + bg - cf - de$$

(12.25)

This is just a few signs different in the imaginary parts from the inner product of the quaternions, (12.18). In particular, we have a swapping of signs in the *ch* & *dg* variables of the first imaginary component, $[1,2]$. We see that the spinor inner product of:

$$IP_{spinors} = \begin{bmatrix} a - \hat{i}b & c - \hat{i}d \end{bmatrix} \begin{bmatrix} \pm(e + \hat{i}f) \\ \pm(g + \hat{i}h) \end{bmatrix} \hat{i} = \quad (12.26)$$

$$ae + bf + cg + dh + i(af - be + ch - dg)$$

matches the anti-quaternion. We therefore have the correlation:

$$\mathbb{H}_{Anti} = \begin{bmatrix} a & b & c & d \\ -b & a & d & -c \\ -c & -d & a & b \\ -d & c & -b & a \end{bmatrix} \simeq \begin{bmatrix} \pm(a + ib) \\ \pm(c + id) \end{bmatrix}_{(\mathbb{H}:\times)}$$

(12.27)

We could have chosen:

$$\begin{bmatrix} -a - \hat{i}b \\ -c - \hat{i}d \end{bmatrix} \quad (12.28)$$

The difference between a quaternion written as a spinor and an anti-quaternion written as a spinor is in the signs of the second component. A quaternion spinor has different signs for the real and imaginary parts of the second component of the spinor; either $a - \hat{i}b$ or $-a + \hat{i}b$. An anti-quaternion has the same signs for the real and imaginary parts of the second component of the spinor; either $a + \hat{i}b$ or $-a - \hat{i}b$. We have:

$$\exp\left(\begin{bmatrix} 0 & b \\ -b & 0 \end{bmatrix}\right) = \begin{bmatrix} \cos b & \sin b \\ -\sin b & \cos b \end{bmatrix}$$

$$\exp\left(\begin{bmatrix} 0 & -b \\ b & 0 \end{bmatrix}\right) = \begin{bmatrix} \cos b & -\sin b \\ \sin b & \cos b \end{bmatrix} \quad (12.29)$$

The difference in sign of the imaginary part translates into clockwise or anti-clockwise rotation. We have two types of spinors differing from each other by the direction of 2-dimensional Euclidean rotation correlated with the two types of quaternions.

### Interim summary:

With proper regard for spinors being conventionally seen as being a Hilbert space rather than a division algebra, it seems fair to say to physicists, "Look, your spinors are really quaternions". A division algebra is mathematically a much tidier thing than a Hilbert space, especially a complex Hilbert space, and division algebras really exist whereas the existence of Hilbert spaces outside of the imagination of mathematicians is questionable.

Given that the only two emergent expectation spaces are our 4-dimensional space-time, from the $A_3$ algebras, and the quaternion expectation space, and given that our 4-dimensional space-time contains classical physics, it seems that the emergent quaternion space might contain quantum physics.

## Bilinear Covariants and the Dirac spinor:

What is a bilinear covariant?[25] What's a Dirac spinor?

A Dirac spinor is a pair of 2-component spinors which is an ordered set of four complex numbers:

$$\text{Dirac Spinor} = \begin{bmatrix} a + \hat{i}b \\ c + \hat{i}d \\ w + \hat{i}x \\ y + \hat{i}z \end{bmatrix} = \psi \qquad (12.30)$$

The Dirac spinor is the variable used in the Dirac equation which describes relativistic fermions (electrons). For fermions (electrons) moving at less than the speed of light (massive fermions), the Dirac equation is a pair of coupled differential equations each with 2-component spinors (quaternions and anti-quaternions perhaps) as a variable. For fermions (neutrinos[26]) moving at the speed of light (massless fermions), the two parts of the Dirac equation uncouple and the Dirac equation with a 4-component spinor becomes two separate 2-component spinor equations.

The adjoint of the Dirac spinor, (12.30), is:

$$\psi^\dagger = \begin{bmatrix} a - \hat{i}b & c - \hat{i}d & -w + \hat{i}x & -y + \hat{i}z \end{bmatrix} (12.31)$$

---

[25] See: Pertti Lounesto: Clifford Algebras and Spinors pg: 137

[26] It seems that neutrinos are not massless, and this throws a spanner into the Dirac equation description of them, but we assume massless neutrinos to fit with the present understanding of the Dirac equation.

## The Spinors of Physicists

The bi-linear covariants of the Dirac spinor are the objects formed from the Dirac spinor, the adjoint of the Dirac spinor, and the gamma matrices that appear in the Dirac equation.

The gamma matrices (also called the Dirac matrices) are, conventionally, the basis vectors of a 16-dimensional Clifford algebra, and the reader might wonder what a 16-dimensional Clifford algebra is doing tangled up with the eight variables of a Dirac spinor, all of which transforms invariantly in a 4-dimensional space-time.

The bi-linear covariants taken together describe the physical state of the electron. The integrals of the bi-linear covariants over space give the expectation values of the physical variables.

The bi-linear covariants taken together form: one scalar & one pseudo-scalar, one vector (four components) & one axial vector (four components), and one anti-symmetric tensor of sixteen components (six independent components).

The scalar is[27]:

$$\Omega_1 = \psi^\dagger \gamma_0 \psi = a^2 + b^2 + c^2 + d^2 + w^2 + x^2 + y^2 + z^2$$

(12.32)

The components of the vector are:

---

[27] Ref: Pertti Lounesto: Clifford Algebras and Spinors pg: 137

## Quaternions

$$J^0 = \psi^\dagger \gamma_0 \gamma^0 \psi = a^2 + b^2 + c^2 + d^2 - w^2 - x^2 - y^2 - z^2$$
$$J^1 = \psi^\dagger \gamma_0 \gamma^1 \psi = 2(-ay - bz - cw - dx)$$
$$J^2 = \psi^\dagger \gamma_0 \gamma^2 \psi = 2(-az + by + cx - dw)$$
$$J^3 = \psi^\dagger \gamma_0 \gamma^3 \psi = 2(-aw - bx + cy + dz)$$
$$(12.33)$$

The six independent components of the anti-symmetric tensor are:

$$S^{01} = \psi^\dagger \gamma_0 \hat{i} \gamma^0 \gamma^1 \psi = 2(az - by + cx - dw)$$
$$S^{02} = \psi^\dagger \gamma_0 \hat{i} \gamma^0 \gamma^2 \psi = 2(-ay - bz + cw + dx) \quad (12.34)$$
$$S^{03} = \psi^\dagger \gamma_0 \hat{i} \gamma^0 \gamma^3 \psi = 2(ax - bw - cz + dy)$$

$$S^{12} = \psi^\dagger \gamma_0 \hat{i} \gamma^0 \gamma^3 \psi = a^2 + b^2 - c^2 - d^2 - w^2 - x^2 + y^2 + z^2$$
$$S^{13} = \psi^\dagger \gamma_0 \hat{i} \gamma^0 \gamma^3 \psi = 2(-ad + bc + wz - xy)$$
$$S^{23} = \psi^\dagger \gamma_0 \hat{i} \gamma^0 \gamma^3 \psi = 2(ac + bd - wy - xz)$$
$$(12.35)$$

The four components of the axial vector are:

$$K^0 = \psi^\dagger \gamma_0 \hat{i} \gamma^0 \gamma^1 \gamma^2 \gamma^3 \gamma_0 \psi = 2(aw + bx + cy + dz)$$
$$K^1 = \psi^\dagger \gamma_0 \hat{i} \gamma^0 \gamma^1 \gamma^2 \gamma^3 \gamma_1 \psi = 2(-ac - bd - wy - xz)$$
$$K^2 = \psi^\dagger \gamma_0 \hat{i} \gamma^0 \gamma^1 \gamma^2 \gamma^3 \gamma_2 \psi = 2(-ad + bc - wz + xy)$$
$$K^3 = \psi^\dagger \gamma_0 \hat{i} \gamma^{0123} \gamma_3 \psi = -a^2 - b^2 + c^2 + d^2 - w^2 - x^2 + y^2 - z^2$$
$$(12.36)$$

The pseudo-scalar is:

$$\Omega_2 = \psi^\dagger \gamma_0 \gamma^0 \gamma^1 \gamma^2 \gamma^3 \psi = 2(-ax\_bw - cz + dy)$$
(12.37)

## Classification of spinors:

Spinors are traditionally classified with reference to the irreducible representations of the Lorentz group; this leads to three types of spinors, Weyl spinors, Majorana spinors, and Dirac spinors. Clifford algebraists classify spinors by their bi-linear covariants[28]. Dirac spinors describe the electron whereas Weyl spinors and Majorana spinors describe the massless neutrino.

Dirac spinors have the scalars, $\Omega_i \neq 0$. Weyl spinors have $\Omega_i = 0$, $S^{\mu\nu} = 0$, and $K^\mu \neq 0$. Majorana spinors have $\Omega_i = 0$, $S^{\mu\nu} \neq 0$, and $K^\mu = 0$.

## Postmethian comments:

Postmetheus was the twin brother of the Greek god, Prometheus, who brought fire to humankind and had a dislike of eagles. Prometheus means fore-thought; Postmetheus means after-thought.

In your author's opinion, the 'pair of complex numbers' notation is not pretty. In your author's opinion, complex linear spaces are a construct of the human mind whereas division algebras really exist.

---

[28] See: Pertti Lounesto: Clifford Algebras and Spinors: pg: 162

Research into rewriting quantum physics using quaternions and anti-quaternions is ongoing. Part of that research concerns the expression:

$$e^{i(kx-\omega t)} = \begin{bmatrix} \cos(kx-\omega t) & \sin(kx-\omega t) \\ -\sin(kx-\omega t) & \cos(kx-\omega t) \end{bmatrix} \quad (12.38)$$

This expression appears very often within quantum physics. We are driven to wonder whether or not we should replace this 2-dimensional Euclidean complex rotation matrix with a quaternion (anti-quaternion) rotation matrix.

Of particular intrigue is the possibility of forming the Dirac equation for particles in our 4-dimensional space-time, massive particles, from two Dirac equations for particles outside of our 4-dimensional space-time, massless particles. It is anticipated that one of the massless Dirac equations will be written a quaternion variable and the other massless Dirac equation will be written with an anti-quaternion variable. The massive Dirac equation might result from the super-imposition of these two equations in a way analogous to how the classical physics field equations, Maxwell's equations, arise from the super-imposition of the $A_3$ field equations.

# Chapter 13

# A Note on Quaternion Fourier Analysis

Traditionally, the sine and cosine functions of the Euclidean complex numbers are used to express functions as a Fourier series. Can the quaternion trigonometric functions be used in a similar way?

Above, (10.20) to (10.28), we have shown that the quaternion trigonometric functions are orthogonal functions by the standard definitions.

Although quaternion space, being a division algebra space with one real axis and three imaginary axes, is a very different space from $\mathbb{R}^4$ space with four real axes, quaternion space does have the Euclidean norm:

$$d^2 = a^2 + b^2 + c^2 + d^2 \qquad (13.1)$$

The orthogonality of the quaternion trigonometric functions together with the Euclidean norm mean that the orthogonal series:

$$FS[f] = \sum_{k=1}^{\infty} \frac{\langle f, \phi_k \rangle}{\|\phi_k\|^2} \phi_k \qquad (13.2)$$

is the Fourier series of the function $f$ over $\phi_i$. Convergence of this series to the given function is not necessarily guaranteed, and the convergence might be of a different nature than uniform convergence.

A function can have different Fourier series using different orthogonal functions. We could use, as is most often done, the two 2-dimensional Euclidean trigonometric functions sine and cosine, or we could use some other set of orthogonal functions, or we could use the quaternion trigonometric functions. Of course, the quaternion trigonometric functions have three variables, but multi-variable Fourier series are well understood.

There is much to Fourier analysis that is beyond the remit of this book. There are questions regarding whether or not the series converges uniformly to the function it represents and much other technical stuff. As far as your author is aware, the quaternion trigonometric functions have not been thoroughly considered as orthogonal functions and as Fourier series by mathematical analysts.

Chapter 14

# Concluding Remarks

We hope the reader is now aware of the most interesting properties of the quaternions. We have shown how to differentiate quaternion functions using non-commutative differentiation, and we have introduced the reader to non-commutative rotation. As far as is known, other than in books by your author, non-commutative differentiation and non-commutative rotation are presented nowhere within the literature. The same is not true of the quaternion trigonometric functions which were first discovered by Hamilton circa 1843. However, Hamilton's notation is a little obscure, and we feel that the matrix notation shows the quaternion trigonometric functions more clearly.

The anti-quaternions are a recent discovery, and, other than in books by your author, are not mentioned in the literature.

We hope we have overturned the 'algebraic extension based on a monic minimal polynomial' view of higher dimensional complex numbers and replaced it with a simpler and thoroughly comprehensive understanding of what division algebras (numbers) are and from where they derive.

We have not delved into the connections between quaternions and Clifford algebras very deeply. A thorough understanding of this connection is presented in the book "The Naked Spinor" by your author. Nor have we delved deeply into the connections to the Lie algebras; we have simply pointed the reader in that direction.

We hope we have demonstrated the double cover nature of quaternion rotation and that the reader has found this 'new' type of rotation very intriguing. We have also presented the recently discovered correct trigonometric part of the inner product of the quaternions.

Of particular interest to the physicist, is the chapter on spinors in which we have discovered two basic types of 2-component spinors. One type of spinor is, allowing for the multiplication operation, algebraically isomorphic to a quaternion, and the other type of spinor is, allowing for the multiplication operation, algebraically isomorphic to an anti-quaternion. We think there is much to be discovered here, and we are intrigued by the possible part that quaternions might play in quantum physics. Mathematically, quantum physics is very messy and quite ugly. Physicists like quantum physics because it works very well; they are prepared to tolerate its messiness because it works so well and are not overly concerned about it not being pretty. None-the-less, it is expected that the grand unified field theory of physics will be pretty, and so prettiness is a thing of value that should be sought by physicists. Prettiness means 'proper mathematics' which is division algebras. It would be a good thing if quantum physics can be constructed from no more than division algebras. Perhaps the reader will be the person to do this.

Your author hopes that the reader has enjoyed reading this book and will, for the remainder of their life, enjoy the knowledge they have gained from this small book. Given that quaternions are rarely taught in western universities,

*Concluding Remarks*

you now understand something which is a little esoteric. You are as a druid among your fellows.

It has been a pleasure writing for you.


Dennis Morris

Port Mulgrave

November 2015

# Other Books by the Same Author

## The Naked Spinor – a Rewrite of Clifford Algebra

Spinors exist in Clifford algebras. In this book, we explore the nature of spinors. This book is an excellent introduction to Clifford algebra.

## Complex Numbers The Higher Dimensional Forms – Spinor Algebra

In this book, we explore the higher dimensional forms of complex numbers. These higher dimensional forms are connected very closely to spinors.

## Upon General Relativity

In this book, we see how 4-dimensional space-time, gravity, and electromagnetism emerge from the spinor algebras. This is an excellent and easy-paced introduction to general relativity.

## From Where Comes the Universe

This is a guide for the lay-person to the physics of empty space.

*Other Books by the Same Author*

# Empty Space is Amazing Stuff – The Special Theory of Relativity

This book deduces the theory of special relativity from the finite groups. It gives a unique insight into the nature of the 2-dimensional space-time of special relativity.

# The Nuts and Bolts of Quantum Mechanics

This is a gentle introduction to quantum mechanics for undergraduates.

# Quaternions

This book pulls together the often separate properties of the quaternions. Non-commutative differentiation is covered as is non-commutative rotation and non-commutative inner products along with the quaternion trigonometric functions.

# The Uniqueness of our Space-time

This book reports the finding that the only two geometric spaces within the finite groups are the two spaces that together form our universe. This is a startling finding. The nature of geometric space is explained alongside the nature of division algebra space, spinor space. This book is a catalogue of the higher dimensional complex numbers up to dimension fifteen.

## Lie Groups and Lie Algebras

This book presents Lie theory from a diametrically different perspective to the usual presentation. This makes the subject much more intuitively obvious and easier to learn. Included is perhaps the clearest and simplest presentation of the true nature of the Lie group $SU(2)$ ever presented.

## The Physics of Empty Space

This book presents a comprehensive understanding of empty space. The presence of 2-dimensional rotations in our 4-dimensional space-time is explained. Also included is a very gentle introduction to non-commutative differentiation. Classical electromagetism is deduced from the quaternions.

## The Electron

This book presents the quantum field theory view of the electron and the neutrino. This view is radically different from the classical view of the electron presented in most schools and colleges. This book gives a very clear exposition of the Dirac equation including the quaternion rewrite of the Dirac equation. This is an excellent introduction to particle physics for students prior to university, during university and after university courses in physics.

*Other Books by the Same Author*

## The Quaternion Dirac Equation

This small book (only 40 pages) presents the quaternion form of the Dirac equation. The neutrino mass problem is solved and we gain an explanation of why neutrinos are left-chiral. Much of the material in this book is drawn from 'The Electron'; this material is presented concisely and inexpensively for students already familiar with QFT.

## An Essay on the Nature of Space-time

This small and inexpensive volume presents a view of the nature of empty space without the detailed mathematics. The expanding universe and dark energy is discussed.

## Elementary Calculus from an Advanced Standpoint

This book rewrite the calculus of the complex numbers in a way that covers all division algebras and makes all continuous complex functions differentiable and integrable. Non-commutative differentiation is covered. Gauge covariant differentiation is covered as is the covariant derivative of general relativity.

## Even Mathematicians and Physicists make Mistakes

This book points out what seems to be several important errors of modern physics and modern mathematics. Errors like the misunderstanding of rotation, the failure to teach the higher dimensional complex numbers in most universities, and the mathematical inconsistency of the Dirac equation and some casual errors are discussed. These errors are set in their historical circumstances and there is discussion about why they happened and the consequences of their happening. There is also an interesting chapter on the nature of mathematical proof within our society, and several famous proofs are discussed (without the details).

## Finite Groups – A Simple Introduction

This book introduces the reader to finite group theory. Many introductory books on finite groups bury the reader in geometrical examples or in other types of groups and lose the central nature of a finite group. This book sticks firmly with the permutation nature of finite groups and elucidates that nature by the extensive use of permutation matrices. Permutation matrices simplify the subject considerably. This book is probably unique in its use of permutation matrices and therefore unique in its simplicity.

# Index

## 2

2-dimensional rotation, 31

## 3

3-dim trigonometric functions, 38
3-dimensional complex numbers, 15

## A

A1 algebras, 29
A2 algebras, 29
A3 division algebras, 26
additive inverses, 14
adjoint of the Dirac spinor, 110
algebraic closure, 10
algebraic extension, 9
algebraic field, 8
algebraic matrix form, 18
anti-commutator, 57
anti-quaternion diff. operator, 63
anti-quaternions, 28, 34, 67, 107
anti-spinor, 108
anti-symmetric variables, 29
axis of rotation, 36

## B

Baker-Campbell-Hausdorff, 84
B-field, 50, 62, 65
bilinear covariant, 110
bi-vector, 71
B-product, 56
B-rotation, 52

## C

Cayley, 11
Cayley numbers, 11
classical electromagnetism, 25
Clifford algebra, 54, 71
Cockle. J., 11
commutation relations, 34, 73
commutation relations, anti-quat., 34
commutator, 57
complex Hilbert space, 98
complex numbers, 9
continuous group, 44, 45
conventional notation, 34
curly-B, 69
curly-E, 69

## D

determinant, 29
differential identity, 69

differentiation operator, 63
Dirac equation, 33, 95, 110
Dirac matrices, 111
Dirac spinor, 110
distance function, space-time, 31
distance functions, A3 algebras, 30
division algebra axioms, 3, 4
division algebras, 3, 16
double cover, 73, 76
double cover rotation, 40, 42, 54

## E

E-field, 50, 65
eigenvalues, 40
eigenvalues, rotation matrix, 37
electromagnetic tensor, 70
electron, 46
electron spin, 45
emergent expectation distance, 30
emergent quaternion dis. func., 31
E-product, 56
E-rotation, 51
Euclidean complex numbers, 9, 10, 14
Euclidean unit circle, defined, 50
even sub-algebra, Clifford, 54
expectation distance function, 30
exponential function, 38, 49, 79

exponential of the matrix, 13

## F

fermion, 95
finite group, 12, 16, 17
Fourier series, 115

## G

gamma matrices, 33
Gauss, 10
Graves. J. T., 11
gravito-electromagnetism, 68
gravity, 68

## H

Hamilton. W. R., 11
Hilbert space, 90
homogeneous Maxwell equations, 70
hyperbolic complex numbers, 11, 13

## I

imaginary numbers, 9
inner product, 85
inner product, quaternion, 89
inner product, spinor, 102
intrinsic spin, 45

## L

left-differentiation, 62
Lie group, 73
linear multiplication, 98

*Index*

local phase variation, 77
Lorentz boost, 13

## M

Majorana spinors, 113
matrix multiplication, 35
matrix notation, 97
Maxwell equations, 68
monic minimum polynomial, 9, 11
multiplicative closure, 20, 22
multiplicative commutativity, 7, 16
multiplicative spinors, 102

## N

NASA, 1
non-commutative differentiation, 58
non-commutative multiplication, 56
non-commutative rotation, 46, 49
non-linear elimination equation, 25
norm, of spinors, 95
numbers, 16

## O

octonians, 11
orthogonality, 84

## P

parity violation, 43

Pauli matrices, 73
Pauli-Schrödinger equation, 95
permutation matrices, 12, 15, 17
polar form of the quaternions, 39

## Q

QFT, 95
quadratic parameter equation, 23
Quantum Field Theory, 77
quantum physics, 96
quaternion algebras, 28
quaternion inner product, 89
quaternion matrix form, 24
quaternion norm, 29
quaternion potential, 58
quaternion trig. identities, 93
quaternion trigonometric functions, 79
quaternions, 17

## R

Riemann geometry, 25
right-differentiation., 62
rotation matrix, 36

## S

scaling parameters, 19
Schrödinger equation, 95
simply connected, 43
singular matrices, 13
SO(3), 73
space-time, 31

special relativity, 13
spinor, 95
straight-B, 69
straight-E, 69
SU(2), 73
SU(2) matrix, 74
SU(3), 78

## T

trigonometric functions, 38

trigonometric identities, 92

## W

weak nuclear force, 77
Weyl spinors, 113

## Z

zero divisors, 7, 13

Printed in Great Britain
by Amazon